Paradox

Paradox

✦

The Rejected Cornerstone

Jane Weir

iUniverse, Inc.
New York Lincoln Shanghai

Paradox
The Rejected Cornerstone

iUniverse books may be ordered through booksellers or by contacting:

iUniverse
2021 Pine Lake Road, Suite 100
Lincoln, NE 68512
www.iuniverse.com
1-800-Authors (1-800-288-4677)

ISBN-13: 978-0-595-34480-2 (pbk)
ISBN-13: 978-0-595-79238-2 (ebk)
ISBN-10: 0-595-34480-1 (pbk)
ISBN-10: 0-595-79238-3 (ebk)

Printed in the United States of America

Contents

Introduction

Socrates once said, "The unexamined life is not worth living." This revered philosopher, perhaps Western philosophy's most famous founding father, fervently believed that, by asking questions and subjecting the answers to logical analysis, it was possible to come to a conclusion about life and morality. His was a unique period in time when elusive, fundamental inquiries could be legitimately ascertained with solid, definitive resolve. Not only did Socrates believe in the human need to question and analyze, but he rendered equally important the necessity for obtaining answers—a far cry from today's growing and foreboding politically correct posturing.

Philosophy, meaning "the love of wisdom," was in its prime. The course of this intellectual pursuit arose from the insatiable curiosity of the ancient Greeks in searching for the facts and laws of nature. Their accomplishments left a legacy of knowledge and art, becoming much of the political and cultural heritage of the West. Statesman Pericles of third century BC not only grasped the significance of his country's contributions to history but also believed that the Greeks had achieved a beautiful and delicate balance in the humanities. He is quoted as saying, "We are lovers of beauty without extravagance and lovers of wisdom without unmanliness."

This golden age could easily be extolled as the exemplary, pivotal point of balance in the history of empires, as the aspirations of mankind for beauty, art, philosophy, and idealism, as well as the symbolism and acquisition of power, reached their peak. Alexander the Great still stands alone as the greatest of all generals—the epitome that Napoleon Bonaparte yearned to emulate over two thousand years later. The establishment of Greek cities, government, culture, and language brought about sweeping changes throughout the expanded empire that represented a new beginning for civilization in a way. This overpowering and highly influential society became the summit and role model for the future, as well as the reflection of the past, for all empires and governments to follow down through the ages. The Roman Empire, for example, absorbed all that was Greek and just romanized it. Then, falling in sync, the Renaissance period, the Classicals, and the Western democratic societies are all basic efforts to reunite with and reestablish their Greek heritage, idealism, and roots, thus spotlighting this remarkable imperial peak in history.

However, what we as individuals or a nation aspire to do or be, compared to what is actually obtained, largely depends on the accepted norm and limitations of the present society in which we live, along with the endowed, accumulated achievements of the past. Like the inquiring Greeks, most of us at some point in our lives, in varying degrees of course, have responded to the formidable task of questioning and trying to understand the true meaning of life by attempting to answer those seemingly unattainable fundamental questions: who, what, when, and why. We have individually and collectively tried to solve the intriguing mysteries of this incredible universe in which we all find ourselves. Unfortunately, our curiosity is usually short-lived and the examination is abandoned because the modern world has generally failed to provide the proper tools to research such lofty endeavors. In spite of our efforts, the process eventually becomes frustrating, and most inquisitors give up the search. Furthermore, when questions continually go unanswered, the overall tendency is to stop asking; our innate curiosity then gradually fades away, giving way to complacency.

The reality of life is, allegorically, like coming into a theater in the middle of a movie, having no clue as to what happened in the beginning or what circumstances led up to the situations now being viewed. But dispelling the mystery is as easy as watching the movie until the end and then staying for the next showing long enough to see the first half in order to view the whole picture. If you want to know the whole truth of the story, it is a simple and logical procedure. If you apply this real solution process to your real life, you will most likely be told that there are no basic and simple guidelines to follow in searching out basic unknowns and that the best you can do is speculate with good "educated" guesses. How would you like to be told by the theater manager that you cannot see the first half of the next show? That you would just have to use your own imagination as to what happened in the beginning, making sense of the story as best you can; then, implying at the same time that you would acquire a feeling of satisfaction just by creating your own fabricated story line to fill in the missing gap. Isn't it amazing that we can accept missing gaps in the realm of truth as a normal and satisfying occurrence but not in the world of entertainment?

Today, as a society we cannot turn to philosophy for answers as ancient Greece did because the scientific world does not recognize philosophy as a "hard" science. We cannot turn to religion, for the logical world does not consider it rational. And, on the last frontier, we cannot even turn to science itself because the world of experts maintains that "true science" can only be applied to the less relevant aspects of our lives, such as math, chemistry, biology, physics, etc., thus making science invalid in the areas of critical thinking, judgmental values, moral issues, religion, sociology, humanities, the arts, philosophy, emotions, and relationships. If Socrates lived in our day, he most likely would not consider life worth living, for there appears to be neither an acceptable way to find answers nor a valid process for examining the main, broader areas of our lives.

Are today's experts right and Socrates wrong? Is it really impossible to answer the age-old question "What is truth?" Is it the mark of an educated person to be content with the assumption that this question is indeed impossible to answer? Through the years, many have devised their own organized and sophisticated views on the mysteries of life. But shouldn't there be more to fundamental truths than just one clever person's ideas or words against another's? Then, on the broader, more general plane, are we all at the mercy of a large segment of educated leaders whose agenda is to create order, saving us from an unacceptable world they think is out of control and void of a basic universal design? Encouragingly, however, this widely embraced rationale is beginning to come under assault from different corners within the scientific and educational community. This relatively new uprising is not only appearing in current books and articles by scientists but is also showing up in popular magazines across the country, boldly challenging old dogmas about design. The following example is an excerpt from a book entitled *The Blank Slate*, by Steven Pinker, featured in *Discover* (October 2002), which finally and refreshingly presents the flip side of the coin:

> Intellectual life today is beset with a great divide. On one side is a militant denial of human nature, a conviction that the mind of a child is a blank slate that is subsequently inscribed by parents and society....At the same time, there is a growing realization that human nature won't go away....An acknowledgment that we humans are a species with a timeless and universal psychology pervades the writings of great political thinkers, and without it we cannot explain the recurring themes of literature, religion, and myth.
>
> As new disciplines such as cognitive science, neuroscience, evolutionary psychology, and behavioral genetics flourished, it became clearer that thinking is a biological process, that the brain is not exempt from the laws of evolution, that the sexes differ above the neck as well as below it, and that people are not psychological clones. Natural selection tends to homogenize a species into a standard design by concentrating the effective genes and winnowing out the ineffective ones. This suggests that the human mind evolved with a universal complex design....Similarly, some anthropologists have returned to an ethnographic record that used to trumpet differences among cultures and have found an astonishingly detailed set of aptitudes and tastes that all cultures have in common. This shared way of thinking, feeling, and living makes all of

humanity look like a single tribe, which the anthropologist Donald Brown of the University of California at Santa Barbara has called the universal people.

None of this is meant to impugn the blank slate as an evil doctrine, any more than a belief in human nature is an evil doctrine. Both are separated by many steps from the evil acts committed under their banners, and they must be evaluated on factual grounds....And the reminder that human nature is the source of our interests and needs as well as our flaws encourages us to examine claims about the mind objectively, without putting a moral thumb on either side of the scale.

Another controversial subject appeared in *Smithsonian* (October 2002) in a book review of *The Genius Within* written by neurosurgeon Frank T. Vertosick Jr. The reviewer is the former *Smithsonian* editor, Paul Trachtman. These authors bring forth significant and valid scientific data as well as common sense in the following presentation that affirmatively undermines the present status quo.

Those who view genes as the irreducible building blocks of life are mistaken....Genes may be likened to rules, the dissenters counter, but a cell may have a mind of its own as it determines how to enact genetic commands. Unraveling a sequence of DNA is akin to learning a word's spelling but not unlocking its meanings in different contexts.

Remarkably, these challenges to the primacy of DNA—an assumption nearly tantamount to dogma—come from the ranks of the scientific community itself, not from creationists or theologians arguing an "intelligent design" of the universe. Vertosick calls into question the gospel according to double helix decoders Watson and Crick, rooted in the Darwinian idea that life evolves through random events as "a blind process, possessing neither insight nor forethought."

I must admit I've waited more than half a lifetime for this book. As a high school student staring for hours through microscopes, I was filled with wonder about the behavior of single-celled organisms, whose life cycles seemed to encompass both randomness and purpose. I perceived sentience in creatures assumed to operate on the basis of instincts and genes alone. But I was taught to dismiss such heresy.

There is mind in nature, he argues, and it's everywhere. Bacteria may not write sonnets, but they have the capacity for intraspecies communication. "Chemistry is their language," he says, "and they've been speaking it for millions of years."

In the final quotation, the author is challenging the longstanding, educational theory that a lack of self-esteem is the root cause of any problem a child may have, educational or otherwise. He doesn't suggest that self-esteem be abolished but that it is dangerous in and of itself. Instead he sees the important need for it to be balanced with its opposite—constructive criticism. The article is a one-page essay written by Andrew Sullivan and entitled "Lacking in Self-Esteem? Good for You!" in *Time*, October 14, 2002.

> You know what self-esteem is: According to decades of psychological and educational theory, it's the essential building block for a successful life. A few generations of children, especially minority kids, have been educated according to the theory that they lack self-esteem, that this deficiency is central to any problems they may have in making their way in the world, and that the worst thing you can ever do to a child is to tell her that she isn't all that.
>
> Well, guess what? Self-esteem isn't all that it's cracked up to be....New research has found that self-esteem can be just as high among D students, drunk drivers, and former presidents from Arkansas as it is among Nobel laureates, nuns, and New York fire fighters. In fact, according to research performed by Brad Bushman of Iowa State University and Roy Baumeister of Case Western Reserve University, people with high self-esteem can engage in far more antisocial behavior than those with low self-worth....Racists, street thugs, and school bullies all polled high on the self-esteem charts....
>
> Self-esteem can also be an educational boomerang. Friends of mine who teach today's college students are constantly complaining about the high self-esteem of their students....Tough professors merely get exhausted trying to bring their students into vague touch with reality.
>
> Of course, in these therapized days, reality can be a touchy subject. It's hard to accept that we may not be the best at something, that we genuinely screwed up, or that low self-esteem can sometimes be fully justified. But maintaining a robust self-image while being able to absorb difficult criticism is surely worth the effort. It could lead to all sorts of strange occurrences: kids working harder, adults exercising self-control, thieves experiencing guilt, even grownup politicians taking full and painful responsibility for their actions and words.

These authors not only share traits of bravery and revolutionary thoughts, but there is also a subtle, underlying theme that parallels and connects their separate examples. Even though their categories were far different, all three discovered something very similar. They each saw structure and design operating on its own, and they all arrived at the same following conclusions:

- It was obvious to them that all the opposing ingredients somehow come together for a compatible and successful result.

- Each one invalidated an either/or decision, choosing instead an equal acceptance of *both* sides to the issue.Notwithstanding and most likely unbeknownst to them, they currently share a common platform. Through their studies and rationale, they have endorsed the validity of duality, defying the accepted norm that has corrupted advanced thinking for decades. At the same time, they have shown the ineffectiveness or the imbalance of isolationism. This designed structure, upon which they landed themselves, is not conjectured or accidental. It is quite legitimate and has been around for a long time, cleverly deceiving us for years, but it is not the enemy.

Thanks to this old, agonizingly perceived foe, a higher range of understanding is surfacing. In our collective minds, this masqueraded villain has been deluding us since time began, baffling its trapped victims in a frustrating stalemate, a dilemma that never goes away. In fact, the greater the education, the greater the checkmate. Disconcertingly, it seems that the more we learn something, the more the opposition grows to oppose it. However, with any acquisition comes its attached opposite in equal proportions, whether we like it or not, such as freedom and crime, blessings and adversities. Ironically, what had appeared to be a per-plexing predicament, a contradiction, and a snare is instead a fundamentally sound equation that provides clear definition, perfect clarity, and the ultimate in organization. Its latent but true identity is slowly being revealed, capturing the elusiveness of truth itself. This intrinsic format exists everywhere and in all things, yet it is institutionally avoided on all fronts. The world continues to search for the universal truth, but the search will be to no avail until seekers finally "face their old enemy head on and look him straight in the eye."

When this happens, it will be quite astounding, because life will never be looked at in the same way again, and mankind can finally move on from this sus-tained mental rut to a higher level of perception. Paradox, the perceived menace to the world of education, has been a stumbling block for years because, even to this day, people continue to reject the true cornerstone of knowledge. They have beaten their heads and "kicked against the pricks" to avoid its pitfalls of entrap-ment, but thankfully it is here to stay because of its essential foundational design and its guided pathway to true understanding. Strangely enough, intellectuals can't seem to live with it or without it, staunchly objecting to the validity of dual opposition, yet, in contrast, "rendering first rate intelligence as holding two con-tradictory ideas in one's head." As this kind of deductive reasoning speaks for itself, it is easy to see that a contradiction on the primary level breeds more con-tradictions. The same way that the first lie requires more lies. Because of an initial error, adjustments never cease, creating a chain reaction of compensations and arguments but never arriving at the fundamental truth.

This book vindicates this marvelous prototype, proving emphatically, once and for all, that a paradox is *not* a contradiction as it so easily appears; this is the deception that has produced irony at its best. Nor is it just a mental exercise to entertain and tickle the psyche. Throughout these pages, the exquisite composition of the paradox is laid out in detail, and its incredible logistical and mathematical laws are disclosed for the first time. Thus, an eye-opening journey into the seemingly mysterious and obscure realm of truth can be embarked upon for real and actually understood by simply observing this awesome, ubiquitous equation, life's universal DNA blueprint. And finally, it is with great hope that the educated hierarchy as well as the average person will come to see that life is not a contradiction as is commonly perceived but is instead a paradox—not an accidental and misfortunate roadblock, but a magnificently complex, creatively designed superhighway.

1

The Search for Truth

Why do the educators of our day seem to be more satisfied with the sophistication of the question and the quest of mystical pondering than with the solidarity of a good answer? In fact, the best way to get along in our society is to be politically correct, which means successfully sidestepping issues or definitive answers. This modern term, *political correctness*, may sound rather nice and polite, but it isn't. What it really means is to avoid decisive labeling deceptively and to erase the line of demarcation between opposites or basics purposely in order to appease or flatter. This may produce peaceful coexistence, but at what price?

In some ways, it is a completely different world now than when Socrates lived, but in other ways it is still the same. The democratic seeds planted so long ago did not take root very quickly. The many freedoms we enjoy today were only in infancy then and rather ineffective for the majority of citizens, yet, in contrast, the "love of wisdom" flourished, thought-provoking ideas were encouraged, and the forum to search out such truths was provided. In comparing the pros and cons of different eras, there is always a tradeoff. Today, the global economy and the new democratic world are progressively erasing trade and cultural barriers, creating greater cooperation and development, but, at the same time, guideposts and boundaries for our everyday lives are slowly eroding away. One of the areas in which this is showing up is the world of expertise: our leaders and educators. This cooperative platform is producing an alarming rate of experts, who fearfully walk the plank of political correctness, with the media eagerly watching for any slip-ups—not to mention the potential power, money, or fame to be gained by avoiding the controversy of division and labels.

Concurrently, when guideposts, boundaries, divisions, etc. are removed, everything becomes one big glob, obscuring vision, clarification, and understanding. Our leaders have opened up our world to more choices and paths to follow, but unfortunately they are taking away our maps and sense of direction along the way. Think how hard it would be to travel across the United States if there were no state boundaries and state signs marking the divisions that give us a sense of where we are. Otherwise, we would be lost. Will this, too, eventually offend someone? Will these lines of distinction be removed as a consequence? If it can be said that "there is no such thing as good or bad," then it may eventually be said that "there is no such thing as New York or California." Without definitive lines of separation, there are no variables or comparatives to enable designations, classifications, or even discernment itself.

This is not to say all experts are bad, nor that all information is garbage. We have all benefited from the experts' expertise, and we can all use all the information we can get. They can and have been very helpful, doing extensive research and experimentation that we don't have the time to do or are incapable of doing. In many ways, they have given us a better life. That, of course, is the good part. However, it is mainly when our assumptions kick in that we begin to get ourselves into trouble. That is the bad part. Most of us can be impressed very easily, and this is what usually creates the illusions. Simply by nature, we are all guilty, at times, of assuming that, if an expert is right in some areas, he or she is also right in other areas. What many do not realize is that these geniuses cross the line of science and speculation quite freely, and, unfortunately, that line of separation is becoming very blurred instead of distinct, with contrasting clarification, as it should be. Most educators mistake the blurring of opposites as creating wholeness, but, as will be proven later, unity and balance cannot be achieved without the initial and essential act of division.

I think many people would be very surprised to know just how much information and sound reasoning is based on "shaky" or completely unknown foundations. Many specialists stack genuine natural processes on top of unproven scientific theories quite frequently, and the unsuspecting masses never question or know the difference. Moreover, many biased experts do not hesitate to build up mounds of sophisticated "scientific" theories without any rudimentary basis whatsoever, with no apologies offered. Some aren't even ashamed to admit that scientists can't find unifying principles that transcend the gritty details, but some have stated that there must be some underlying simplicity and order to this thing called life.

A good example of these missing gap theories is in the foundational law of gravity. Astronomers have learned a great deal about our universe and have accomplished some extraordinary things. However, they are quick to tell us that they do not know what gravity is even though it is the single most basic fundamental force that rules and balances all heavenly bodies as well as the earth. If you apply this reasoning to "real life," would anyone in his or her right mind construct a physical project without a proper knowledge of basic foundations? Of course not, how scientific, or wise, would that be? Nevertheless, we tend to trust the experts' views, based on what they know, and ignore the bigger picture as to how little they know.

Without sufficient curiosity as well as confidence, we have collectively let the experts rob us of our own analytical powers, mainly because we lack the proper tools to question the validity of their findings. They have offered us a world of complex sophistication that we find difficult to comprehend, and yet they have failed to give us simple basic underlining principles that we are capable of learning. Thus, the majority of those living in this modern society are beginning to find themselves stuck in the middle between two worlds: expertise and basic knowledge, neither one being easily attainable. One of my favorite things to do is research children's encyclopedias and science books because the writers are forced to get down to basic causes and effects and actually state them. I suppose political correctness is excused or considered harmless at this level.

Thus, we are a society void of good individual judgment at a time when it is needed the most, for we are bombarded every day with new items, new ideas, new controversies, and new avenues to follow. There are thousands upon thousands of experts constantly giving us advice. They tell us what to do, how to be successful, how to be rich, what to buy, what to believe, how to vote, how to be beautiful or sexy, how to be healthy, how to be happy, how to be marriage partners, how to raise children, to spank or not to spank, and so forth. Accordingly, when there is so much to choose from, it makes the task extremely difficult, especially when the mechanics of good decision making are slowly eroding away and becoming virtually unknown. This present-day generation is experiencing an enormous paradox. On the one hand, they have an unprecedented freedom of choices, both physical and mental, yet on the other they are the least equipped in learning how to make choices.

The Bombardment of Information

If you are an average person trying to learn and understand life the best you can, your own in particular, how do you know what is best or what is right, and how do you live your life accordingly? Should you believe what the experts say? If so, which expert, since so many disagree with each other? How do you decide what is right for you or wrong for you, what is false or bad, what is true or good? This bombardment of specialized expertise and seemingly endless choices has intimidated the public to a no-contest match, causing most of us to fall into three general categories at some time or another. The first group, which is a growing segment, is getting just enough education to be highly impressed and will follow along, shifting to whichever wind is blowing hardest at the time, tenaciously following the popular trend. The second group simply stays confused most of the time, and readily admits it, or goes along with the current flow rather unconsciously. The third group is totally out of the jet stream, balking at everything, and readily boasts of their biased opinions about everything. They adamantly stick with their own narrow viewpoint and are proud of it because nothing else makes sense or interests them. It is, perhaps, their only defense against something they cannot understand because it seems beyond their reach. The good side is that they may not be taken in by a lot of the trash, but unfortunately they are unable to go forward by taking advantage of the good stuff.

Against such odds, how do we, the average citizens, go about this difficult task of discernment and learn for ourselves what is important and what is right? How do we compete with the dictatorship of experts? How do we fight back and what should be our line of defense? How do we learn to be confident, qualified, independent thinkers, worthy to control our own thoughts and destiny? How do we bridge this enormous gap between king and servant, between specialized expert and naive consumer? And, finally, how do we counterattack the false assumption that the experts are always right, or that their methods are untouchable and above us all? How do we do this, since there is not a single-minded, all knowing, totally unprejudiced, flawless person on this earth we can all turn to? Perhaps that is good. When the new colonists in America rejected their king, they did not turn to another king; they fought back with individual sovereignty and intelligence.

It is "we, the people" who need to learn how to discern wisely between all these choices and make good judgments. This is something we must and should do for ourselves and not leave to others. But do we dare question the great arena of experts? Do we fight, or do we bow down to the kings? When there is good and bad information, plus too much information, someone needs to sort it all out. There is not another human being on the face of earth that is more positioned or more responsible for taking the facts from the experts and custom tailoring them to your own wants and needs than you. No one else knows you better than yourself. But the question is, how do you become a good and reasonable judge? How do you trust yourself? Is there a viable way to stand up to these mental giants yet, at the same time, take advantage of what they have accomplished, applying it to our lives or throwing it out? Could there possibly be guideposts out there that the experts are ignoring or destroying that can help us answer these basic questions on our own?

Could there possibly be a formula that precisely measures all things in the same simple way that we know with certainty that two plus two equals four? Are there basic fundamental principles or formulas on which all knowledge and the pursuit of it are based? Isn't there any way we can prove, mathematically, that what we consider to be true, which others may consider false, is really correct other than "It's my word against yours"? If there isn't, then how do we know we are right and they are wrong, or whether they are right and we are wrong? The bottom line is that there is no way we can know with certainty unless there are proper mechanical and fixed laws for the learning process and, consequently, for our understanding. Therefore, do such procedures exist? And, if they do, are they user-friendly? Or are they reserved for the privileged few, only those endowed with genius-like capabilities? Is there a course somewhere or natural guidelines that show us exactly how to go about properly defining, labeling, and governing our thought processes, a simple and specific measuring tool that anyone can use? And, ultimately, if that is the case, shouldn't we be able to measure truth and validity in all matters in the same precise way that we solve simple math problems?

These inquiries lead to other important examinations. Can mankind even perceive a correct and organized view of life if it is not fundamentally organized in the first place? If this hypothesis is true, is logic possible in a chaotic world? Can we have reasonable and organized thoughts if nature is not already reasonable and orderly? Which came first, logic or design? Can they perhaps be separated? Is mankind organizing a corrupt and random universe, or is a designed universe creating organized minds? Or, to the contrary, does disorder and order exist at the same time? If so, how do we know the difference? Just what is the true definition of order or disorder? Finally, is there a process by which we can legitimately and confidently answer these difficult questions, thereby knowing with certainty what is truth?

Laying Foundations

Before building a foundation in the quest to find and establish true knowledge, we must begin at the very bottom of the ladder by questioning and examining everything, making sure the labels and tools we are working with are essentially correct. As in real life, when building anything new, we must start from scratch. This can only be done by breaking the confines of our common assumptions, by starting fresh and opening up our minds the best we can, for man does not think in a vacuum. Real understanding takes time and experience. Until that happens, we have stockpiled many first impressions that are quite distorted and fabricated as well as ingrained. Consequently, they are not easy to get rid of. In fact, it is more like major surgery. The restructuring of thoughts and opinions is a very arduous process, plus the emotional let down of knowing we may have been wrong can be hard to take or can even lead us astray.

There is always a price to pay when making a change. The more that is involved when making that change, such as reputations, prestige, responsibilities, income, etc., the higher is the cost. Therefore, it shouldn't be surprising to anyone who understands human nature that the normal thing to do is to follow the path of least resistance by changing and risking as little as possible. Consequently, the higher the position one has, with so much at stake, the least likely change is going to happen on its own. So it would be a safe bet to wager that the top experts in all the many specialized fields are the least likely candidates to lay it all on the line, opening up to question and change. As a result, it would be a fair assumption to say that the greater the expert, the greater the closure. However, what we value the most will weigh heavily in our quest for the truth.

Once the decision is made to open up our minds and start fresh, we must establish our basic understanding of exactly what terms such as logic, science, and law mean, thereby determining the basic building blocks and tools for learning. Furthermore, and of utmost importance, we should never assume anything, but we should question everything. The only way this can be done is to examine properly the transactions of nature and our thoughts in their most basic form, thus becoming aware of what actually constitutes laws and assurances. Upon such investigations, the inescapable conclusion will inevitably be that confidence can only come through consistency. Transactions that occur over and over again, without fail, become predictable, and the things we can predict we call laws. A law is an automatic and predictable response, a fixed consequence of certain actions. Like the default mode designed in a computer system, it is automatic. This is the one thing that you would think all the experts would agree upon. Therefore, what separates laws from speculations is a consistent arrangement or design of things and events forming patterns that occur every time. Once we can use the terms "all" and "every time" in our collection of facts, we have ourselves a law. Thus, keeping minds open and looking for consistent patterns is the beginning process that leads to real answers.

The dictionary's definition of logic is as follows: 1. The science that deals with the rules of sound thinking and proof by reasoning; 2. Connection, as of facts or events, in a way that seems reasonable. The definition of science is "a branch of study concerned with observing and classifying facts and attempting to make up general laws about them." The only sound conclusion that can be drawn from these definitions is that reasoning and logical thinking are derived from observing, associating, and classifying data of a fixed and organized subject. Without a doubt, for us to be able to classify knowledge systematically, which is science, there has to be order and regularity in the subject we are studying. Besides a capable, orderly, and receptive brain, the main source of logic comes from the consistent patterns in our environment. It is a prerequisite for whether our thinking can be reasonable or not.

If there were no order and predictability in our surroundings, there could only exist confusion and unpredictability in our minds as well. For example, if gravity worked only on occasion, there would be no such thing as the law of gravity. It would be something we could not explain, count on, or perhaps even know. Think how insane it would be if all the consistent things we rely on daily, consciously and unconsciously, suddenly became disarrayed and unpredictable. Therefore, our logic is inseparable from our environment. We are able to acquire sound scientific knowledge only because sound and consistent laws do exist. Consequently, for us to know anything, we must look for, comprehend, and analyze consistency in our environment. Without order, design, and predictability already existing, we could not have logic, and, without logic, we would not know anything for sure. Hence, there would be no laws to rely on mentally and soundly.

Furthermore, if logic is dependent and applicable to the lessons we have learned from nature, it would be ridiculous for us to try and reason apart from the exact laws and understanding we have derived from our surroundings. Yet, inherently, this is what mankind has a tendency to do all the time. However, our logic is really not logical if conclusions are unsubstantiated by the existing designs we find in our environment. If reasoning does not have a sound fundamental basis upon which to reason, why try to rationalize? Thus, logic and consistent order (design) go hand in hand. Predictability, because of the consistency of the elements and their laws, enables us to reach sound conclusions with the reliance and confidence we need. No matter how smart we may think we are, our brains are not enough to create sound and reasonable minds; it requires both. Therefore, true logic can only be obtained if proper, systematic, cerebral functions (internal) are substantiated by pre-existing and predictable, external laws.

Unfortunately, scientists who mix speculations with classified facts have given science a bad name. They have made it hard for society to sort out the difference between their interesting but speculative theories and true science. They prey on people's naive assumptions. Science programs that stray from true, established laws and that attempt to address those "who, what, where, or why" issues always leave their audiences hanging between their highly educated guesses to the abrupt absurdity that mankind will never know such answers because these kinds of questions cannot be measured mathematically. The typical lecture or narration will impress upon its viewers some highly specialized information, but, in the last few minutes, the pattern is invariably to cancel out all reliability on what was just demonstrated because "We cannot be absolutely sure about anything." The contradiction disappointingly leaves an empty and lost feeling, with an eternal uncertainty looming over the conclusion instead of satisfaction, solidarity, and confidence building.

Is this the correct mindset, to not know satisfactorily, with certainty, the most meaningful things in life yet eagerly stockpile an endless array of facts and trivia that have no basic fundamental connections whatsoever? Is this what we have to settle for? As mentioned in the beginning, is true science confined only to the less important aspects of our lives, such as math, chemistry, biology, or physics but invalid in the areas of critical thinking, judgmental values, moral issues, religion, sociology, humanities, the arts, philosophy, emotions, and relationships? Is logical thinking only logical when applied to certain areas and illogical when applied to others? Wouldn't it seem reasonable that logical thinking is logical regardless of where it is applied? Isn't it unreasonable to think that logic is sometimes logical and sometimes not? Either there are exact and reliable basic laws designed in the learning process, period, across the board, or there aren't. It would also seem rational to think that, if reasoning relies on subjects that have fixed laws, then the reverse would have to be true as well. Consistent external laws would require that the mental perception of these laws also be based on consistency and on a fundamentally sound process in order to be properly interpreted.

If, on the other hand, there are no consistent, predictable, and underlining universal laws in the learning experience, in which rules come and go by someone who decides where they should be placed or where they cannot, then we are all in trouble. For, if that were the case, there would be no definitive right or wrong, good or bad, and the best of all foggy notions about discernment would be no better than the famous line by the philosopher Descartes: "I think, therefore I exist." Does this mean that a rock, which cannot think, doesn't exist? What does that statement really mean anyway? What do we really learn from it? Nothing concrete. It keeps the mind in a state of confusion. Nor is anything to be achieved from his mystical notion that we cannot determine whether we are asleep or awake. These highly subjective thoughts just tickle the imaginary psyche with sophisticated appeal, leaving thoughts dangling, suspended in midair, and totally ungrounded. Just from these small examples, it is no wonder why the field of philosophy is not listed in the accepted journals of science, making it difficult as to which section to put it in, fiction or nonfiction.

Therefore, if foundational questions about how we think and learn cannot be answered logically and decisively with good, sound mathematical principles, then how can we confidently say or know that what we think is right is right and what we think is wrong really is wrong? If there are no such definitive laws, clear definitions, or lines of demarcation, then anyone's speculations are as good as anyone else's, and the question would be "why bother?" In the practical world, would anyone in his or her right mind consider building bridges and buildings without solid and decisive measurements? Of course not! Neither would anyone send his or her child to high school without learning the elementary basics of reading, writing, and math. Then why do a lot of professors and scientists build and state facts on shaky, unknown, underlining principles, or basic building blocks, and get away with it? On the other hand, if there are specific laws that govern and measure the way we think, how things are legitimately determined in our minds, then why haven't the experts found them?

It has always been mind-boggling to me how oblivious brilliant minds can be when it comes to making connections. Case in point, there isn't a single unit or transaction anywhere that exists or functions without the support of an equally opposing agenda. Yet, the scientific community still hasn't made a generalization of this phenomenal consistency. For instance, the terms "basic" and "general" are exact opposites that operate in equal and opposite directions, specifically in response and proportion to each other. The more basic the category, the more general it will be; the lower you go down the ladder, the broader will be its outward extension. It is automatic.

For example, it is a geographical fact that Americans live in the United States; however, the subjects become more basic and broad, at the same time, when it can be said that human beings live on earth. Moving the pendulum in any direction will immediately increase or decrease the opposite end accordingly and proportionally. This basic exercise is as scientific, mathematical, and predictable as Isaac Newton's law that "For every force, there is an equal and opposite force." Yet, astonishingly, it appears to draw no attention to the scientific community of that fact.

If we are true students and observers, confidently relying on these consistent laws that apply down to the basic core of nature, the next step would be to let that automatic swing take us on an educational journey abroad towards all categories, making the unique connection that exists between basic and general, simple and complex, as well as one and all. Consequently, if it can be proven that simple bottom-line laws which dictate the basic learning experience do exist, then that one set of fundamental laws would coherently as well as systematically spread abroad and apply to all learning situations and all ideas, no matter what the subject.

New Discoveries

Failure to find a unifying universal law frustrated Albert Einstein, even until the time of his death. Because he could see a certain amount of chaos existing within order, more or less as a flaw, he could not see it existing systematically outside that parameter as its equal and legitimate partner, therefore he made the famous statement: "I can't believe God plays dice with the universe." However, if he had lived long enough to witness the chaos scientists' incredible discoveries on the duality of order and chaos, he might have had regrets that he hadn't taken his law of relativity far enough into broader and deeper dimensions. The relationship and laws that exist between mass and energy are the same for order and chaos. They are, as he discovered, different forms of the same thing and interchangeable, which is exactly the same as saying these two opposites are basically the same thing except in reverse. I believe that, if he were living today, he would agree with scientist Joseph Ford, who afterwards said, "God does play dice with the universe, but they're loaded dice."

The new science of chaos has been slow to catch on, for even the authors who presented this revolutionary idea to the public several years ago realized the reluctance they would be facing. The following is a quote from the book simply titled *Chaos*, by James Gleick: "Each scientist had a private constellation of intellectual parents. Each scientist has his own picture of the landscape of ideas and each picture is limited in its own way. Knowledge was imperfect. Scientists were biased by the customs of their disciplines or by the accidental paths of their own educations. The scientific world can be surprisingly finite....No one misunderstood the discovery of the molecular structure of DNA, for example. But the history of ideas is not always so neat."

When you learn what chaos really is, thanks to the chaos scientists, you will surprisingly discover it to be order on a higher level than mankind is normally used to. After you look past the so-called chaos and tap into this very advanced system, out of it will emerge patterns of the kind of order we can recognize. Chaos has all the simplicity we can relate to, but the simple patterns are not apparent until after we have studied the complexity of a system, which is its main characteristic. It is not until we learn something about this complex organization that we begin to recognize some familiarity and common order hidden inside.

Those who only see apparent contradictions, confusion, and, mostly, randomness in the world we live in as well as the world that is in us blind themselves to the beauty and predictability of this highly sophisticated order. And, to the contrary, those who do see a lot of order, the kind we are used to, but regard the rest as disorder, as Einstein did, fail to see the depth and grandeur of the more advanced and complex side of our world (i.e., chaos).

This relatively new and seemingly unchallenged science appeared in the movie *Jurassic Park* for a brief moment. The scene portrayed it more as an entertaining, mind-boggling mental exercise rather than a highly organized, plausibly understood foundational law of nature, having the exact same pattern and relation to order as matter has to energy. The following quote shows that the new science offered more than mere entertainment: "Chaos poses problems that defy accepted ways of working in science. It makes strong claims about the universal behavior of complexity. The first chaos theorists, the scientists who set the discipline in motion, shared certain sensibilities. They had an eye for pattern, especially pattern that appeared on different scales at the same time. They had a taste for randomness and complexity, for jagged edges and sudden leaps....The most passionate advocates of the science go so far as to say that twentieth-century science will be remembered for just three things: relativity, quantum mechanics and chaos."

Without the aide of the computer, these discoveries may not have taken place. However, by putting these theories through a computerized long-range weather test, one of the scientists, Edward Lorenz, observed "an image of predictability giving way to pure randomness." But Lorenz saw more than randomness embedded in his weather model. He saw a fine geometrical structure, order masquerading as randomness. The simplest systems are now seen to create extraordinarily difficult problems of predictability. Yet order arises spontaneously in those systems—chaos and order together. Only a new kind of science could begin to cross the great gulf between knowledge of what one thing does—one water molecule, one cell of heart tissue, one neuron—and what millions of them (all) do." Because of the earth-shattering results of mass and energy's exchanges, it is quite understandable why these elements receive so much attention, whereas structure, design, and order take more of a back seat in human interests. However, these two sets of opposites, mass/energy and order/chaos, fit the same basic set of rules, in which their two halves simply represent each other and are totally interchangeable. Since they share the same unique design and laws of relativity, they should be given the same respect. These elemental compounds are some of the most basic categories known to mankind. Nevertheless, in the proceeding chapters it will be proven that there are other categories, more foundational than these two, which also reveal the same predictable design and pattern of exchange.

As far as these relatively new data discoveries concerning nature's innately designed arrangement causing a complete flip-flop in the way order is perceived, it is respected by some people. However, most are not able to make such an abrupt turnaround and recognize that disorder is really a higher form of order. Nevertheless, computer programs driven by mathematical equations from non-linear dynamics have created a new scientific field that reveals a secret inner order to what has appeared up to now to be chaotic phenomena. This may sound intriguing and poetic in print, but in all practicality how does one make the giant leap from what we normally know as chaotic to sublime? How do we transfer this into reality when the word "chaos" automatically congers up an ugly vision of disorder? It has always represented an undesirable state of affairs and rightly so. However, there is a lot more to chaos than what's on the mere surface.

A Reality Check

The unconscious mind and the involuntary functions of our body have proven to be higher forms of order and intelligence than on the conscious level. Without a doubt, the unconscious brain has significantly more data than the conscious brain because it has permanently stored all previous information and calculations in its files, whereas the capacity of the state of consciousness is quite limited, as well as temporary. Furthermore, the sophistication and enormous tasks of the body's involuntary actions are far greater than our own voluntary efforts. Educational science films demonstrate graphically how the human body's internal "armies" make governmental armies pale in comparison to their number, efficiency, power, and strategic complexity. Also revealing and, perhaps, a little disturbing is the abundance of invisible microorganisms that cover our entire body, providing a very useful and organized mission—to rid it of dead skin and unwanted organisms. In fact, they claim these little creatures clean the body better than we do with all our cleansing chemicals, which, if used too excessively, inhibit these minuscule mites from during their job well.

Order and chaos, conscious and unconscious, voluntary and involuntary are all dual categories that help manage our lives. One of the nice things scientists are discovering these days is the necessity for both; the more balanced they are, the better off we are. It was recently revealed that athletes perform better when their conscious, voluntary actions of stress and intense concentration are matched with an equal amount of serenity. The reason is that, when the element of fun, relaxation, or confidence is added to the mix, the unconscious brain comes into play by tapping into all information previously calculated and stored from years of training and practice, along with any naturally gifted abilities inherited through the genes. Thus, the athlete would be getting the best of both worlds, conscious and unconscious, when it is balanced of course. Too much mental concentration evidently squashes out the unconscious (inherent) side and vice versa. I know a great high school coach who instilled in his baseball players basic skills, strategies of the game, and a great intensity to win but balanced it all with tremendous praise, encouragement, and the insistence of having fun at the same time. Needless to say, he brought out the best in his players and was a winning coach.

Another scientific study found that, during the sleep cycle, the brain used this time period to sort out and properly assess information taken in during the previous conscious hours, a much-needed service. Therefore, a lack of sleep could be detrimental to our mental state as well as our physical state. The same balancing act of opposites is needed in our own daily encounters with order and chaos. When either side dominates, we get less than satisfactory results. Consciously taking on too much responsibility and order can not only be too stressful, but it can also shut the door to others or other elements that could be very beneficial. On the other hand, not enough hands-on management in our lives allows too little control, which brings us to the flip side of chaos.

In our personal lives, the unfavorable concept of disorder is needed as a backdrop to arrange things properly the way we may want or need them, but, in reality, when something is out of our control, then that disorder falls into the control of someone else or something else. Beyond our personal use of it, it is really not disorder at all, but just order that is not ours anymore. If that dispensation is returned to nature, such as a neglected backyard, then the inevitable fallout is absorbed into a much higher performing, large-scale, and recycling ecosystem. What appears on the surface to be chaotic disorder is instead very complex transactions that are far beyond our own limited capabilities. It may be momentarily unpleasant for our immediate situation, but its involvement is concerned with a much greater, more encompassing, and highly needed balancing agenda.

There are many other examples that follow the same laws of inevitability. If we don't control our budgets, our assets are taken over by larger financial institutions. If we don't structure or discipline our children, social workers will place them in someone else's care. If we don't check our morals, higher institutions that employ prison guards could monitor us. If the sovereignty of a country is not managed well, it may be taken over by a government that is managed well. If we do not take care of our health, we will become fodder for an unseen but immensely larger, more systemized population than ours, and so forth. The bottom line is that it is good for us to control what we should and, at the same time, see the beneficial need for a higher, complex order that is not under our jurisdiction. This kind of reality check helps us balance our lives between existing opposites plus achieve higher levels of perception, knowing without a doubt that what we do not properly manage will be absorbed by a more powerful, highly organized, and sophisticated system we call chaos, which may not be wanted but, overall, is vitally needed. Moreover, it is essential that we understand it is not something inferior and without order, after all, as it so easily appears, but is in fact quite the opposite: organization on a grander scale.

Organization at Its Best

There is an old saying: "An answer is only as good as the question." Therefore, in order to obtain a good answer, it must be matched by an equally good question. Few question the logic of that statement, yet fewer still, if any, see it as a mathematical equation in the form of an automatic and balanced structure. Nor do many see the same parity in Newton's law of forces—for every force there is an equal and opposite force—or its connection to Einstein's law of relativity in which mass and energy are equal opposites, being the same thing except in reverse. Neither are many impressed by the geometrical similarities in order and chaos. Why is that? One of the reasons is because of the way we normally look for things. How many times have we overlooked our missing keys when they were in plain sight the whole time? The answer is that we usually find things if we are consciously looking for them in a particular spot or area. We have been taught not to look to the mathematical field to solve our normal, everyday, personal activities or problems.

One of the most important questions anyone could ask is the question hardly anyone considers: what is the best form of organization, order at its best, and where do you find it? Does that system or arrangement already exist? If so, where does one look? To the experts or to nature? Are we still waiting for a genius to finally set up the ultimate standard? It is the one question, if answered correctly, that will surprisingly answer all basic questions. The following chapters are designed to mathematically prove that the fundamentals of good organization are neither something humans make up, nor are they created by the best mind the world has to offer. They will, shockingly, show that the nature of our world and universe is already uniquely structured, and it is only when we mimic its orderly, formatted transactions that we too get the same progressive results. The pleasure and power we get today from modern technology are because mankind correctly studied nature and successfully copied it, not invented it.

There are three basic learning stages. The first and most elementary stage is pure speculation, in which information is never grounded, never forming true consistent patterns, with no basis or evidence for one's ideas. The second stage is a little better but extremely dangerous. These thoughts recognize predictable patterns but impatiently interject shortsighted and biased opinions into the process before it has completed itself. A half-truth can sometimes be worse than no truth at all and, as the old saying goes, "a little knowledge is a dangerous thing." However, the third represents the ultimate stage because this process never strays from the complete path of its learning field. This path is not an experimental, misguided one waiting for some genius to charter its course but is of the same format upon which all things are formulated and measured.

Fortunately, there is a perfect scale for organizational excellence already existing, a place where anyone can go for true answers. The vast stretch of seemingly endless variations that are produced between two opposite poles is the perfect, organized, learning range and pathway towards real understanding. Anything outside that range is disconnected and theoretical. Ironically, this concept is not foreign to the present world of education. In fact, educators use this great expansion and comparative tool all the time in their teachings, but, unfortunately, they fail to see its profoundly fundamental value outside the classroom. It takes on many forms: a continuum, a spectrum, a paradox, polarization, duality, opposites, yin and yang, and so forth. However, the astounding difference is that they perceive these arrangements, for the most part, as mystical problems to avoid or as contradictions, which is the extreme opposite of what they basically are—distinct, valid, unique, and universally designed. It is the same injustice afforded the ultimate deception: when good is called evil and evil, good.

Once, when reading an introductory college textbook, I paid close attention to the authors' explanation of a continuum: "a range of possible alternatives between two logical extremes." A half-circle graphic was illustrated with two labeled opposites at the ends, with the middle, or the balance of the two, halfway at its peak. They proceeded to explain that "the continuum was the main conceptual device and comparative tool used in this book to survey the field of philosophy" with the following precaution, "there is nothing philosophically sacred or ultimately true about the continuum concept. In fact, some philosophers argue that understanding is distorted or precluded when ideas are viewed as totally opposed to one another." Being quite stunned and knocked out of focus by such a conflicting statement, I temporarily stopped reading, began flipping through the entire book, and, sure enough, caught a glimpse of at least two or three continuums in each chapter as their main teaching tool and standard of measurement. Needless to say, I was stunned by the contradiction. That was the beginning of my real education into the world of contradictions and a rollercoaster ride into irony.

Furthering my studies on the continuum, I found it to be, conversely, a hidden treasure of understanding waiting for the benefactor that discovers its amazing structure. It is perfection, completion, education, and true organization all rolled into one, giving true meaning to the term "pathfinder," for its dual format is intrinsically set up as an informative and comparative tool. When thoughts and studies are housed and collected within a well-organized system, it provides proper instructive guidance, creating confidence in what is being learned. The paradoxical continuum is that perfect system because it is connectivity and consistency at its premium as well as organization at its best.

Staying attached to a continuum, following its guiding path, keeps the mind from straying off into obscurity and speculation. Following the true course of knowledge and disciplining our biases is extremely difficult but possible and necessary if we are to be successful in our pursuit of true understanding. Staying within the confines of a continuum and maintaining the incredible balancing act of its two opposite bookends requires extreme effort and focus, but, if maintained, it is the true form of perfection. In the following chapter, it will be proven how these simple, down-to-earth key words *balance* and *perfect*—normally subjects victimized by surrealistic, religious, or mystical interpretations—are fundamentally and definitively one and the same when placed in an informative springboard of two opposites.

2

The Balancing Act

The terms *balance* and *perfect* are good examples of subjects that spin off into clouds of obscurity when they become unattached to their continuums. These words normally conjure up all kinds of speculative thoughts in the public's mind. But again, when we come down and connect to the real world by dealing with their practical aspects, the true meanings become quite clear. For instance, it is so easy to think when things are still or calm that nothing is going on, but in reality the complete opposite is true. When something is perfectly balanced in nature, it means equal and opposite forces are at work, ironically creating the illusion of inactivity. Each side neutralizes, not erases, the other side; furthermore, the greater the balance, the greater are the opposing forces and thus the greater the deception.

Therefore, balance is not a condition that is stagnant and devoid of energy, contrasts, or oppositions, as it would seem, but it is an active mathematical equation. If a person knows anything about math, he or she will understand that an equation is a mathematical statement expressing the equality of two quantities, as the dictionary defines it. Consequently, the balancing act requires two opponents that are equally divided. Yet, if you tell the average person that you cannot have balance or calm without division, strife, and opposition, he or she will most likely disagree with you. The same misconception is also applied to the word "perfect." Perfection represents the pivotal point in nature, including human nature, that marks the exact spot between equally balanced and opposing positions. Contrary to common opinion, it is not an isolated, unapproachable purity that is without precise measurements of something we know.

Initially, my opinion was no different. In preliminary studies on the paradox years ago, I took my research into the field of philosophy and, not surprising now, ended up in an intoxicating, ironic stupor. Instead of escaping the mind-boggling dilemma of paradoxes, I found that the greater the mental stretch, the greater was the paradox and thus the greater the irony. It turned out that the main division that separates the great Western philosophies from the great Eastern philosophies is in the concept of paradoxical duality. However, what I found to be such a strange and incredible twist of consequences was that it is indeed accepted by both sides, but only in reverse applications. Paradoxically, one side considers it ideological but not applicable, and the other applicable but not ideological.

As mentioned earlier, the West applies the paradoxical continuum in its studies but invalidates its authenticity, whereas I found it to be the exact opposite in the East. While studying the ancient philosophies of the East, I discovered that it was much more difficult separating religion from pure philosophy than it was with the West, but, when I searched as far back as the *Tao*, I settled upon what I saw as the summit of Eastern philosophy. Its concepts, more or less, trickled down into all Eastern philosophies as well as Eastern religions. When I began reading statements such as the following, I immediately thought to myself: "Now this is more like it!" "The Taoists saw all changes in nature as manifestations of the dynamic interplay between the polar opposites yin and yang, and thus they came to believe that any pair of opposites constitutes a polar relationship where each of the two poles is dynamically linked to the other."

But then, as in all roller-coaster rides, there eventually comes a time when what goes up must come down, and my escalated enthusiasm began to plummet to the ground when I read the following passage that summed up a kind of Taoist paradise. "Men might be possessed of the faculty of knowledge, but they had no occasion for its use. This was what is called the state of perfect unity. At this time, there was no action on the part of anyone—but a constant manifestation of spontaneity. The Taoist sage does not strive for the good but rather tries to maintain a dynamic balance between good and bad." Therefore, the ironic, gyrating message ricocheting in my head came through loud and clear from the East and from the West, and that is what works for them ideologically doesn't work operationally. Disappointingly, each side practices the opposite of what it believes. To sum it up, one side is proposing the absurdity of obtaining a balance without the struggle and the other, a misguided struggle without the intent of balance.

In practical terms, an interim balance of any kind cannot be achieved without an extension, or active force, into opposite directions. You cannot arrive at a balance by doing nothing; you do it by extending and equalizing the two opposite dimensions. Further, you achieve stability in the middle by the interactions of the opposing ends, similar to a circus tightrope walker. Nor can you have both weights on one side: one has to oppose the other (across the line). For example, a pair of scales is unbalanced until equal weights are placed on both ends opposite each other. Neither could a graceful, gravity-defying ballet leap be maintained if there was not a really strong effort made to resist the downward pull. The art of making something look easy, when, in fact, it is very difficult, takes tremendous forces on many levels. Furthermore, can you imagine how wasteful it would be to a weight lifter, after years of training and bodybuilding, if gravity were to disappear suddenly? With nothing to massively resist, muscles would eventually decay, depleting the accomplished physique.

From the Western viewpoint, what good is the struggle of resistance if there is no striving towards or even the recognition that the coming together of opposites orchestrates harmony and unity? This single factor alone may account for the reason the West emerged as a rather stress-oriented but industrially progressive society and the East as a peace-oriented but passive race in a rather regressive cultural state—instead of the recognition and harmony of the two oppositions. However, in the long run, I eventually saw these two opposing extremes as a large, balanced continuum of its own, each being the exact opposite of the other, making them philosophy's "perfect" bookends.

Balance is perhaps the principle ingredient and crowning glory of the paradox formula, for it represents the holding force and harmony of its opposing elements. The equilibrium, resting in the middle between equal opposites as the ideal state of being, is not really a new or revolutionary idea in itself, only in the degree that it has been conceptualized. There have always been bits and pieces to this insight, but they have remained isolated because of the reluctance or inability to carry the idea over into broader and deeper categories, as well as the insurmountable task of being accepted by a large segment of people.

The dictionary accurately defines balance as "stability produced by even distribution of weight on each side of the vertical axis—equality between the totals of the two sides of an account—an aesthetically pleasing integration of elements, harmony." Without contrasts, diversions, or oppositions, there would be no such thing as perfection, precision, or harmony, for there would be nothing to bring into balance as well as nothing on "the other side of the axis." Nor can you integrate something that is not initially divided in the first place. Being the successful state of anything, balance is hopefully the end result. Therefore, if a person seeks to have success in a particular area of his or her life, no matter what the category, no matter how small or large, a balance must be achieved. And in order to have a successful life, he or she must bring all areas of life into an overall balance. Consequently, when the divided opposites within a unit are completely balanced, harmony prevails.

There is a perfectly good reason why some stories, plays, or novels are considered great and others are not. The qualifying standard would not require a fifty-page wrangling essay but just a simple and concise mathematical equation instead. A great story is basically an integrated balance of extreme emotional and circumstantial opposites. Both sides of the equation have to be engaged equally for success and perfection. An overextended margin from either direction beyond the halfway mark would leave an undesirable or unbalanced result. For instance, the worse the villain, the greater is the hero; the bigger the suspense, the greater is the mystery; and the worse the tragedy, the greater is the love story. In order to be right on target, they have to match up exactly. The following is a list of other successful balances:

1. What makes a good game?   equally matched oppositions, not one-sided

2. What makes a good painting?   a unique and balanced blend of many contrasts

3. What makes a good account?   a balance sheet

4. What makes a good ecosystem?   a balanced environment

5. What makes a good court case?   equal representation from both oppositions

6. What is perfect vision?   20/20

7. What makes a good marriage?   equal partners

8. What constitutes good health?   a balanced diet, exercise in moderation, enough sleep but not too much, neither all play and no work nor all work and no play, motivation yet easy on the stress, emotional stability and government stability, plus the cooperation of a balanced "Mother Nature."

Thus, mankind is dependent on all the balances of the above, plus many more, in order to live a balanced life, making it obvious that neither side of the equation is removed or eliminated but just monitored and balanced. Simply stated, whatever is on one side, there has to be the equal amount on the other side for a legitimate balance.

It is hard to imagine anyone with a fair amount of education who is not impressed with the eternal and dramatic pursuit of nature to maintain a balance. This tireless activity accounts for all energies that exist in us and in our universe. Every unit, out of its own attractions, will always fight or react to maintain a balance of its components unless separations are imposed upon it by stronger outside units; then it becomes part of an even larger and more ultimate balance. Substances and forces are always shifting to fill in or subside in an adjusting ebb and flow effort to achieve balances. The atoms themselves present a harmonious balance of opposites, with every nuclear proton corresponding to an orbital electron so that the overall charge is neutral. Negative ions attract free electrons to balance electric charge. Radioactive isotopes decay toward stable atomic forms.

The existence in nature of imbalance leads to actions designed to produce the opposite—balance. The greater instability, the greater is the impetus toward stability (a balance in itself). The greater the expansion of opposites, the more extremely delicate is the balance and, unfortunately, the greater is the vulnerability. When the early twentieth-century scientists descended into the basic fabric of life by manipulating its arrangement, the impact was broader. For instance, having surpassed the atom bomb by going a step lower, the largest destructive blast known to man was discovered lying dormant in the tranquil state of a small nucleus. When you finally get to the one, you will find the all. All the many battles engaged in by the multiple and varied systems of this world to attract or repel, creating balances and imbalances, despite the sometimes random and chaotic aura that appears to exist, are pulled together and held in check by the boundary of an overall ultimate unit. The numbers all add up to one incredible equation divided by two. Philosophically, logically, physiologically, scientifically, or mathematically, it couldn't figure any other way.

Like It or Not, Duality Exists

Duality, the stumbling block and rejected cornerstone of modern education, is, ironically, the key ingredient to understanding and must be vindicated before any great strides in educational reform can be achieved. The blurry lines that exist and have existed for a long time between contradictions and paradoxes, between division and wholeness, between black and white, must be clarified and strengthened if we are ever going to rightly divide the word of truth. Incredibly, this is the point in higher education where the experts are hung up. They apparently cannot understand how the world can be divided and whole at the same time; therefore, most reject the concept of opposites and deny it any respectability or validity, treating it like a contradiction.

One of the most common phrases on the subject of opposites is the following: "I believe there is more to life than just black and white." This is absolutely correct, but what people do not seem to realize is that, because of these two bookends, we have a seemingly endless continuum arrayed with beautiful colors for us to experience and enjoy; there's a lot more to black and white than what meets the eye. We cannot see the hidden colors in white when they are all reflected as a whole. Nor can we see all the hidden colors that are absorbed into black objects.

However, this does not mean they are not there. It is easy to assume that white and black are void of color, which couldn't be further from the truth. In fact, it is just the opposite: they each contain *all* the colors but in reverse applications. Consequently, it is the varied pluses and minuses of these two poles that create the many spectacular colors we see.

Another reason for the reluctance to accept the paradoxical concept is the emotional side. Naturally, it is hard to think about, look at, and of course experience the dark side of the equation, much less admit the frustration of trying to eliminate something that doesn't go away but, in fact, increases as things get better. As things change, so do the trade-offs, but the balanced ratio of opposites stay the same. World leaders remain baffled by the attachment freedom brings, which is crime, and, obviously, they do not see the automatic correlation between the two. Mainstream sociologists naively think they can increase freedom without increasing crime and refuse to face the inescapable conclusion that, if they want to reduce crime, they must also reduce freedom proportionately. The central message emanating from any society is an old and brutal paradox: good and evil are inseparably entwined. Freedom is the spirit and nucleus of Western culture as well as the brilliance of all its grandeur, but it is the root of social injustice, crime, and greed. Air travel is a good example of this process happening today. Seeing policemen or military personnel in our airports enforcing strict securities, creating a type of military state, would have been inconceivable a few years ago before the terror issue of 9/11 came home to roost.

The flip side to good, which is bad, inevitably keeps rearing its ugly head and reinventing itself where and when we least expect it. It would be less painful if we could have balance and harmony without opposition, love without hate, freedom without crime, peace without war, good without evil, and pleasure without pain, but that is not the way it is. Why is this? Most experts see this as being caught in the midst of a contradiction. Many wonder if this is an accident of nature or part of a botched plan. However, the more sophisticated and advanced we become in our rationale, the more we realize there is a reason for this automatic flip-flop arrangement. This is perhaps why many experts are beginning to use the word "paradox" more often in their rhetoric, even though they still haven't accepted it yet nor see a perfectly good reason for it.

The old phrase "You can't have it both ways" is another way of saying you have to accept the whole, two-sided package that comes with a choice of action, meaning you cannot pick one side without reaping the inevitable consequence of the other side. For instance, the old saying "You can't have your cake and eat it too" simply means that if you want to enjoy eating your cake, then, of course, you cannot keep it. The alternative is that, if you want to save the cake because it's too pretty to consume, then you simply could not enjoy eating it. Most of us can understand this little riddle; however, when the same deductive reasoning is applied to other levels, this simple logic goes out the window. It is a very difficult mental exercise to always see duality in a situation, but it must be achieved if one is to have a balanced viewpoint.

"There are two sides to everything" is a common phrase, but few really believe it. In a news magazine there was a story about a man who had weathered Hurricane Iniki after surviving the 1989 Loma Prieta earthquake and losing his house to the 1991 Oakland fire. In my first natural reaction, I immediately thought to myself, this either is the unluckiest man alive or else he is the luckiest man alive. The usual initial mindset is that it has to be one or the other. Which is it? Then I correctly saw it as a perfect paradox, because he was both, each side being dependent on the other. He could not have been the unluckiest man if he had not been the luckiest as well and vice versa, as the following demonstrates:

1.   This situation consists of two totally opposite viewpoints:

   • He was unlucky for three awful disasters to happen to him in such a short period of time.

   • He was lucky to have escaped these disasters with his life three times.

2.   Each requires the other side to exist in order to receive its own identity:

   • He could not have been the luckiest man if the unlucky disasters had not happened.

   • He could not have been the unluckiest man if he didn't have the good luck of surviving with his precious life in order to experience the next disaster.

3.   Both sides are emphatically equal, of the same intensity:

   • How devastating it can be to lose valuable and precious possessions and witness horrifying natural disasters (the bad).

   • He was shaken profoundly into the reality of how insignificant material things are and how precious life is (the good).

The tragic death of Princess Diana reveals another "two-way street" dilemma that everyone tries to avoid. Her life alone has proven that, if you want fame or media attention for a worthy or selfish cause, you will automatically reap the unwanted attachment that it brings. With the invasion of the news media come notoriety, distortion, and harassment as well. Naturally, everyone seeks a favorable coverage without intrusion and ridicule, but it doesn't work that way. Besides good and bad, true and false, fame and privacy also work in proportion to each other. The more the fame, the less privacy there is, thus creating the dangerous situation that led to her death. Ironically, however, the downside of being blasted all over the tabloids was the very thing that created devotion from the masses at the time of her tragedy. In a world where few people know their neighbors well, millions were peeking into her personal life through some of the more disgusting acts of journalism, and, as a result, they couldn't get enough of her glamorous fairy-tale lifestyle. Whether the image was real or false, it didn't matter. The whole world, it seems, had a heartfelt attachment to Princess Diana because of the way the paradoxical process works.

Some of the transactions of this subtle structure are recognized, but, for the most part and to the general public, it is totally misunderstood and unappreciated. In a strange twist of circumstances, our highly cultured, literate world puts this magnificent and unique formula into practice every single day, but, ironically and blindly, it fails to see it in action, thereby denying the perfection of its truth as well as the beauty of its mechanics. What educated mind doesn't appreciate the intellectual benefits from understanding the theory of relativity, which is a tremendous comparative tool for measuring what we know? What physical therapist or fitness expert would advocate doing away with opposition because resistance and the struggle it requires is the only way to build up body strength? What news media would dare give a biased report by not showing the dilemma of two sides being ironically right and wrong at the same time? And, finally, what reputable self-help consultant, guiding his or her clients towards achieving ultimate success in the category in which they are seeking help, would point them in any other direction than to reach a balance?

Therefore, what the experts are obliviously practicing and blindly relating to us is that life experiences must be presented in a balanced forum of two sides that can, ironically, be seen with a measure of truth in both, totally opposed to one another at the same time, yet resulting in equal relativity. The thing that houses all these ingredients is, of course, the paradox, which instead congers up confusion and frustration in most people because it is so completely misunderstood even though it is a part of their everyday lives. If these laws were understood and exercised in their own lives, they would achieve the same thing the paradox does. Not only would they have a wide range of options clearly defined the way it does, but this simple little network would provide the yardstick for monitoring their actions between two basic poles, making it possible to achieve a balance, the true definition of success.

Consequently, the educational world should stop denying duality as if it were an accidental, frustrating dilemma or contradiction. Hopefully, educators will soon come to recognize that the paradox is not only the basic framework that sustains the universe but is also the very same organizational tool we can use to measure and monitor our own lives instead of letting them be dictated by others. Its guidelines can give us analytical powers beyond the experts and beyond our wildest dreams, thereby bringing excitement, enlightenment, and control into our lives as well. Regardless of its bad rap as contradictory and mystical, it is the universal underlining principle unifying all things that all the great minds are looking for but can't seem to find. It is the simple mathematical equation that defies uncertainty. It is the perfect scientific cause and effect principle. It is the exact science that can be exercised in all the important personal aspects of our lives and not just the impersonal ones. And, in spite of its political incorrectness in the ranks of professional diplomacy, it can be properly and clearly defined.

Dissecting the Paradox

The word paradox not only presents a concept that is, at first, puzzling to comprehend but rather tricky to define as well. The dictionary describes it as "a statement seemingly absurd or self-contradictory, but really founded on truth." On the other hand, "self-contradictory," representing two conflicting ideas that cannot exist together, is a deduction. Deduction comes from the word "deduct," which means to subtract. Logically, that means that, when two ideas are in complete conflict, one must be eliminated in order to be reasonable. Likewise, in a contradiction one side automatically voids the other because, if one side is true, the other cannot be. It is total rejection. A paradox is not a contradiction because neither side of a paradox can be eliminated. Rather, it is an adduction. Adduction comes from the word "adduce," which means to bring to, such as add. Rationally, this means that one side or idea adds proof to the opposing side. They match up and coexist, verifying each other's existence.

Para in Greek means "beside, near by, and along with," in more practical terms, "two." And the Greek meaning for *doxa* is "opinion." Therefore, the logical adduction of the two words would be an opinion that has two sides. Instead of a contradiction, the paradox is a valid two-sided adduction, providing clarification, not confusion. Consequently, the only things befitting the description of something that seems to be a contradiction but isn't and comes in pairs are opposites. They are the only things that always exist in pairs. As will be proven, what appears to be the most opposite in life is in reality the most similar. Opposites form a rather unique combination, because nothing is more divided than opposites, yet nothing is more common. They are an inseparable pair. Cold is obviously not the opposite of white; just as hot is not the opposite of black. Cold and hot, as well as black and white, belong side by side, paradoxically linked, like all opposites, locked together yet divided in a self-contained unit. The reason there is nothing more similar than an opposite is because opposites are the same thing except in reverse. And the reason they are divided is because they work separately, in opposition to each other but connected in interdependently supportive transactions.

Opposites are really just paradoxes but represented in simpler form. The same laws that apply to a paradox apply to opposites as well. Being simpler, it is easier to get down to the heart and soul of this back and forth process. It is such a foreign concept to most people because it is opposite to the way they normally think. However, if one studies the laws of opposites and sees how they actually function in real-time experiences, it is much easier to understand. Looking at the world through bifocal lenses is hard in the beginning, but, as with anything, it gets easier with practice, and it is necessary to acquire a balanced viewpoint. Even though you may have the correct template, you still have to size it up in each new category, doing the homework and continuing to combat old ingrained ideas that are hard to change.

Therefore, the next phase of examination is to go further into the paradox and examine its two basic fundamentals: opposites. First, their basic laws will be laid out and then examples as to how opposites actually work together will be demonstrated. The aim is to portray fully how the basic, beautiful, rhythmic, pulsating, and reverse reactions of the paradoxical unit are uniquely designed and how they work undercover in everything and in our everyday lives. As mentioned earlier, when something works well, the mechanics go undetected. When the weather is so pleasant, the tremendous forces that create that serenity are invisible and unapparent. In many other examples (e.g., a performance, a dinner party, a business meeting, etc.), the recipients are unaware of the hard work and planning that went into making these endeavors a success. In fact, the easier it looks, the better the results are and, notwithstanding, the harder it was to pull it off.

Case in point, I remember a few years ago a woman in the news who was being divorced by her husband for a younger woman after many years of marriage. She in turn sued him for half the financial proceeds of his very successful business. He claimed she was just a housewife, never doing anything to deserve half his company. She rebutted by saying he was so busy with his work away from home that she practically raised the children by herself, being the one who was there for the homework, to attend all their functions, settle all the squabbles, do all the disciplining, and see to all their personal needs. She not only managed the household, but she often entertained his clients with extravagant dinner parties at home as well. Through her hard work and ingenuity, she took the domestic load off him as well as helped land many of his business deals through her entertaining, thereby playing a major role in the success of his company. Then putting her finger right on the nerve, she said, "the reason it appeared to him that I didn't do anything was because I did it all so well and ran things so smoothly that it made it look to him that I didn't do anything." Here again, things are never the way they so easily appear.

Trying to match wits with an older, very philosophical brother, I began early in my teenage years inspecting and analyzing natural phenomena. Even then, when I started noticing the duality of nature, there seemed to be two of everything. Not knowing what to do with this material, it became a hobby and a pastime to look for opposites in day-to-day processes. I noticed that, if I couldn't find an opposite to something, then I had at least found a slight variation of it, leading me to my first observational law. By creating different mental scenarios, I began to understand that opposing variations were the beginning stages of awareness, a prerequisite to learning.

Consequently, I invented the following hypothetical scenario in order to learn the most basic, elemental stages of understanding. I conjectured that, if everything we could see were similar to an endless, white, blank sheet of paper, with no wrinkles, no shadowing, or variations whatsoever, it would be the same as being blind, even though we could physically see it, thus making it clear that just seeing illuminated matter is not enough to make one see. I reasoned it had to be accompanied by its opposite for it to be comprehended; the opposite contrast enables the eye to discern and differentiate objects. Then I envisioned upon this imaginary white sheet of paper that a black spot appears for the first time, thereby creating a learning experience. With the help of this contrasting spot, the awareness process began. Not only is the color black observed, but now, for the first time, its opposite, white, is noticed as well. This perception required both at the same time.

Next, I theorized that upon that same white-paper scene the exposure was to light gray instead of black. Even with this slight hint of black, white is noticed for the first time. Then, upon further learning stages by experiencing darker grays, I realized that we would never learn the full potential of how white white is until we have learned how black black is. In fact, white never becomes its whitest until it is set next to jet black. This introduced me to the next law of opposites: the learning stage is conditional to an equal and opposite ratio. Just as Newton's law states that for every force there is an equal and opposite force, it is equally true that the more we learn of one opposite, the more we learn in equal proportion of the other. In conclusion, I learned that, if we cannot find an opposite to something, we have at least found a slight variation to it; otherwise, we wouldn't even be aware of its existence in the first place.

I also experimented with the other senses and got the same results: they all worked according to the same laws. But the next hypothesis was to me even more interesting and revealing. I envisioned a world in which all matter had the same degree of softness, no variations whatsoever, in which everything felt exactly the same. Bringing into play the previous examinations, I reasoned that we would not only be ignorant as to what hardness is, since it's not in the picture, but the awareness that our environment was soft would never enter our minds. Then I continued in this imaginary situation with a person having touched 15,000 items and another person having only touched 1,000 items, but the very next thing the latter person touches is hard. Then I realized that the 14,000 extra items the first person touched didn't bring him any closer to the truth than did the one that experienced an opposite. These hypothetical examinations made it quite clear to me that the acquisition of knowledge not only requires associations but that the basis for those associations has to be opposites.

The most basic and initial learning experiences naturally begin in infancy, making it impossible for us to remember how humbly it all started. Among the newborn's first contacts in life is being held by the warm, soft, cuddly arms of its mother, and the infant consequently begins to associate these sensory feelings to the easing of hunger pangs from her breast milk. Not only is he learning what hunger and fulfillment, warm and cold, soft and hard are, he is also developing mental pictures (opinions) from these associations, such as pain and pleasure, fear and security, likes and dislikes (the beginning stages of love and hate). As these physical experiences advance, so do the mental concepts. Small adductions made from the repeated associations of mother's presence and milk (2), plus warmth and comfort (2), equal fulfillment (4), and lead to advanced deductions such as love and confidence. Advanced opinions are all based upon the foundation of these elementary associations of opposites, typical of building blocks; one must precede the other and, therefore, takes considerable time and numerous learning experiences.

These early examinations helped me realize how truly limited and humble humanity is, that our mentality cannot go beyond our experiences, what we have been given. We can mix, match, manipulate, and rearrange all possible combinations of what we already know and have, but we cannot even bring to mind what we have not yet experienced. Neither are we responsible for the "bookends." Nor does one opposite create the other: both exist simultaneously. However, once we have the polar ends of a continuum, knowledge can be expanded with all possible arrangements of these two poles. Most people mistake these new combinations as something entirely new and inventive. I can't help but recall King Solomon's wisdom in this matter: "The thing that has been is that which shall be and that which is done is that which shall be done: for there is no new thing under the sun." (Ecclesiastes 1:9)

The following mental exercise was done in fun during my youthful, philosophical exploits and I used it a lot on my unfortunate family and friends, but no one seemed to appreciate it like I did. The subject is "nothing." I would declare that there was no such thing as nothing, and, of course, they would all argue with me. I may have entertained myself with this debate over the years, but the logic has stood the test of time. It is essentially true that we are not capable of imagining something we have not yet experienced. When it is said, "There is nothing there," what we are really saying is that what we know to exist or to have existed is no longer here or in a certain locality. It is a concise way of saying that some thing or things we have learned and previously experienced are presently absent. Besides, if you can think of something and call it nothing, that makes it something, not nothing. And on the other side of the coin, remember just because we can't see, hear, smell, or feel anything doesn't mean something is not there; in fact, it's quite the opposite. There are more things present that we do not perceive than that we do. In all the widespread spectrums of nature—energy, light, sound, smell, etc.—our sensory contacts occupy a very small portion of those very large ranges. Consequently, what is comprehended through our senses, compared to what abounds, is minute. Hence, our world is mostly made up of things that people normally call nothing.

Through my own experiences and by observing others, I noticed a peculiar pattern in human nature: the more knowledgeable exposure we have, the more aware we are of how little we know. For instance, the more we know, the more we realize how much we didn't know. To the contrary, the less we know, the more we seem to think we know. Consequently, the different degrees of awareness affect opinions. How many times have we thought there wasn't much to some thing or to a particular subject until we turned our attention to it and discovered there is a whole world to be learned there? Each new door opens up new fields and dimensions of knowledge that we were previously unaware of, new angles to look at; thus, there is no end to learning. Yet the less we knew, the less we thought was there or the more we thought we knew.

Moving forward in age and rationale, I eventually realized the basic element to life wasn't two but one, divisible by two equal opposites. Obviously, the first order of research is simply to narrow a particular subject down to its least denominator, one distinct category, then to divide it between its two opposite bookends, which always exist, and then you complete the process. I reasoned that, since there is one of everything, all things can be broken down to one single digit, a specific category, or a unit, thus one bottom line. Until I got down to the ground floor, there could not be a beginning foundation, and, without that foundation, there could not be a proper structure. However, the labeling is just the beginning format. Until that single unit can become productive, it has to divide itself equally into two equal halves, just like cells that divide and multiply equally. Then and only then do the awareness and learning process begin.

In time, I began to see the learning process in the same light as simply putting a puzzle together. When a certain piece fits its designated slot, you know it belongs, because the slot, the receptacle, is the exact and reverse opposite of the projectile. There is no guesswork on our part; either it fits perfectly the slot it was designed from, or it doesn't. Nor does it require a genius, just an average mind that is willing to put the patterns and pieces together laboriously. The following principles are a recapitulation of the basic laws of opposites that unfolded after years of observation and research, which I continue to use as a template or guideline to measure all things.

- All units (categories) have two basic opposites: everything has an opposite.

- Opposites are found equally coexisting: one does not produce the other.

- For everything, there is an equal and opposite opponent.

- Nothing is more divided than opposites, yet nothing is more similar.

- Each is a carbon copy of the other, the exact same thing except in reverse.

- Each opposite is the sole source, reason, and provider of the other's identity.

- The more that is learned about one opposite, the more that is learned about the other, equally and automatically.

- Each opposite has a set of major and minor, negative and positive, components.

- All the interactions of opposites operate in reverse form.

- What is dominant in one is recessive in the other and vice versa.

- The dominant traits are designed to give and the recessive traits to receive.

- Each unit is complete within itself, both giving and receiving, yet is a dependent half of a larger outer unit, creating independency and dependency all at the same time.

Studying the structure and inner workings of opposites helps tremendously in understanding the seemingly unfathomable paradox. In return, understanding the paradox unlocks the mysteries of the learning process. Knowing the basic rules, stated above, as to how we simply learn anything, no matter what, will help lay a solid foundation of confidence, because the systematic consistency that is seen in the smallest of transactions will miraculously be seen in the largest of systems. To the contrary, success towards understanding our thoughts is impeded if the fundamental laws of nature's dual characteristics still remain vague and invalid in our minds.

By examining and relying on these scientifically sound transactions, it will be discovered that the organization for one is the same as the organization for all. Therefore, in conclusion, whether it be life in general (all), or anything specific in this life (one), if the basic skeletal outline of its composition is revealed in the smallest of its parts, then, obviously, the DNA has been found.

3

All in One and One in All

Upon further examinations and analysis, the overwhelming evidence will clearly reveal that a paradox, a continuum, and a pair of opposites are all basically one in the same. Each has a structural framework, plus mechanical transactions meeting the same criteria, whether it is used in tracking a spectrum or categorizing two simple opposites such as black and white. As mentioned earlier, the majority of educators today still do not endorse the idealism of opposites. However, they continue to use the continuum in their classrooms on a regular basis as a practical and decisive measuring tool. Regardless of their idealistic protests, a continuum is, in practice and in principle, two interacting equal opposites forming a unifying and clarifying system. They may not know why, but educators apparently use the continuum in their studies because nothing else can provide an informative and comparative layout the way it does.

Not only is the continuum the platform upon which all understanding should be based, but there is one continuum that represents, by its own unique design, the model format for organization. Again, it isn't a foreign concept. Rather, it is actually stated in a very common, everyday phrase, but, irrespectively, this "DNA jewel" isn't taken very seriously. Laying aside for a moment the scientific investigation into the learning process, in practical terms, what could be more conclusive or perfectly complete than to have what you are studying or organizing in an all in one and one in all arrangement? Even at the human level of creation, this is the orderly process used when something is organized well: anything from a simple toy box to sophisticated computer software, from a Cub Scout meeting to a UN assembly. In fact, what could be better organized than "a place for everything and everything in its place"? Or what could be more loyal than "all for one and one for all"?

These popular axioms are more than just neat catchy phrases. Just to say they are more is a tremendous understatement, for they essentially embody the true and perfect form of organization, the "mother" of all continuums. Besides being equally balanced opposites, one of the reasons this organizing unit is perfect is because it's complete. It (one) includes everything (all) and everything (all) is contained in and based on one. There are some people who do see the validity and parity in the proverbs stated above, how they apply to real-life situations, but, apparently, that is all they see. This is unfortunate, for they are missing the "jack-pot" of understanding and a well of information found in this little phrase. Its subtle design contains, without a doubt, the key to the mysteries of life, the platform upon which all things were created, and the paradigm whereby all things function.

Before anyone should consider these bold statements as utterly absurd and politically incorrect, it must be realized that laws or solidarity of any kind could not be established if absolutism, which this organizing tool represents, did not exist. What if Isaac Newton made a statement saying that "For almost every force, there is an equal and opposite force"? Or what if Albert Einstein had said that "In most cases, mass turns into energy and energy into mass"? What if the force of gravity attracted some matter, but not all? If that were the case, these exact laws that have helped mankind establish a foundation in knowledge and create technical advances would be nonexistent. Unless all things being organized, classified, or studied are included, the arrangement would be fragmented and incomplete as well as unscientific. For example, all matter is made up of molecules, all molecules are made up of elements, and all elements are made up of atoms. The consistent theme throughout these laws is that nothing is left out; there are no exceptions. Pulling it all together under one roof for a thorough research is the only way to enable a true and genuine examination. Finally and ironically, all these vital laws were founded on one concept: absolutism, the taboo in modern society.

Under the Microscope

In order to decipher this intriguing all in one and one in all continuum, its basic design must be dissected and studied following the dictates of its embedded, instructional guidelines. First and foremost, it is obviously an equation divided into two opposite halves, which are the exact same thing except in reverse. All in one is the same thing as one in all, just turned around. Each side contains all the variables, being equal to one another. Additionally, not only are the two sides equally opposed in their arrangements, but each side is subdivided into two opposites as well. All is the exact opposite of one, just as parts are precisely opposite to the whole. All represents the parts or contents, that which is contained, and one represents the whole, the container of all parts. In the initial procedure, one is the organized system that houses all things: "a place for everything." Then, in the second half, all is all things individually housed in that setup: "everything in its place."

There is a very important feature to note in observing these polarized units. Each of the two sets has a dominant and a recessive trait but in reverse order, whatever is dominant on one side will automatically become recessive on the other side and vice versa. Therefore, one half's dominant trait matches up with the opposite half's recessive trait and vice versa, thereby creating the magnetic attraction (positive and negative) needed to cross the line of division in the initial base equation, completing a larger whole unit. A basic law of dominant and recessive roles is size. Remembering this simple rule will help tremendously later when working this equation in new areas. In the continuum, the dominant trait is always the larger, elongated, or elaborated one, whereas the recessive trait is always condensed and withdrawn.

As the first half, all in one, suggests, the major emphasis is on the whole, one—making it the dominant, elaborated trait—with minor emphasis on the parts, all—making them the recessive, condensed trait. This side of the equation is mainly viewed as a group, a system, or an arrangement and not so much for its contents. For example, a human body is made up of many diverse parts all working together to create one unique person. When focusing on that person as a whole, you do not necessarily notice these parts that all come together producing this end result. If you do, it takes away from the whole concept. One name and one person, with all his or her many infinitesimal parts and features, should all blend together as one, representing one personality.

The dominant goal and objective of all in one is collecting all things in a category under one roof and lining them up systematically to create a whole. Ironically, this adding process, to seek out all possible items, putting them in their proper place and leaving nothing out, is aided by this side's recessive partner by no other means than subtraction: in order to increase, you must first decrease. This is not surprising, however, when one realizes that all transactions and laws of motion are accomplished by reverse form, in which all components are interdependent on opposite support systems. By diminishing the differences and accentuating the similarities, it enables a summarization of behavioral patterns, or consistent laws, into one basic and general cognitive description abstracted from all the parts for a screening device, thus allowing the recognition of its own for the sake of consolidation and boundaries. The augmentation and glorification of the whole is very much dependent on the humble and recessive role of the parts.

In the second half of the equation, the dominant goal and objective is to do just the opposite of the first half: diminish the similarities and accentuate the differences. In one in all, the major emphasis is switched to the parts, all, consequently making them the dominant trait, with minor emphasis on the whole, one, now becoming the recessive trait. What was once condensed (all) is now specialized and prominent. Instead of unity, the main focus is on individuality and uniqueness of the varied parts. All parts share the one basic profile, but, when it comes to details, no two are exactly alike because each one is a repeated but varied version of all the others. By elaborating on each individual part, it provides a tremendously exciting learning field with its seemingly endless variations to observe, each one adding new information and characteristics to the equation that none of the other parts can. On the other hand, what was predominantly important in the first half (one) is drastically reduced in the second half in the form of an abstraction, which is intrinsically embedded, like DNA, in each and every enhanced part. This subtle blueprint creates a solid foundation for keeping the differing, glorified parts safely anchored and connected to its innate system of order (one) as well as providing the essential value of relativity.

In summary, even though these back and forth transactions can be rather confusing, their primary concepts and guidelines are quite simple. It is as easy as doing basic math, such as division, adding, subtracting, multiplying, and transposing or understanding the elementary principles of magnetism. There is a scientific reason why the equation is divided into two's as well as their halves. These two components in each case are opposites, which are basically the same thing except in reverse. This transposing of magnetic fields and their attraction of opposites is what generates the activity between the dual parts. Not only are the subjects, one and all, opposite to each other, but their mathematical actions are as well. When one is reduced, the other is being added and vice versa. Furthermore, each act and focus requires total dependency and cooperation of the other; one cannot increase unless the other decreases and vice versa. Each opposite needs the other to accomplish its own duties.

What an amazing comparative tool for consistency, understanding, and authenticity when the whole envelops all the parts and the parts reveal the whole. Mathematically, everything (all) in a category is a part of the whole (one), and the whole (one) is represented by all its many, varied parts (all). "Just do the math." After setting up the equally divided equation, you subtract, then you add, and vice versa. In one half, you primarily generalize; in the other, you mainly specialize. It is one very simple organizing process, but, at the same time, it incorporates within its confines a multitude of crisscrossing transactions in order to complete the whole gamut from the least, one, to the greatest, all, hence making it very complicated, with practically no end to the possible variations within a classification. Everyone keeps looking for infinity somewhere out there when, ironically, it is in here.

Equations in Action

When you think of extremely complicated systems that produce great works and perfect harmony, such as an accomplished symphony orchestra, it is hard to envision that they are based solely on a basic equation of just two equal sides, but that is the uniqueness of the perfect continuum. Not only are musical sounds a result of the zigzagging, vibrating, opposite motions of electromagnetic waves and the back and forth movements of the conductor's hands or wand, but also incorporated within the confines of an overall orchestral continuum are innumerable, fundamentally contrasting equations, such as percussion/wind, performers/audience, orchestra/conductor, etc., that are being produced and multiplied; all are transacting the same way they are being encased, much like images that are reflected between two mirrors facing opposite each other. In this process, each mirror is reflecting the opposite mirror, with lower levels of the exact same images being repeated over and over again at slightly different angles and sizes in a seemingly endless, varied display until it drops out of sight. The two opposite bookends are where the whole orderly arrangement begins and ends, but, within that duo system, what starts out as very simple becomes very complex. Great achievements and successes don't just happen in a conglomeration of circumstantial happenings under the mystifying surrealism of a genius. They happen because someone either purposely or obliviously followed the design of the perfect structure.

By following the precepts of the all in one and one in all continuum, all studies can be organized in a wonderful forum of control and consistency, giving us the kind of confidence and real understanding we've all been looking for. It enables the confidence to know that all things in a category (one) are collected, assembled, and balanced within its own framework. Once we have the two opposite bookends, the least (one) and the greatest (all), then it is just a matter of doing the homework by filling in the middle, which is basically their own vast array of offspring. It leaves out the guesswork, because it is an organized spectrum having specific guidelines and boundaries to follow. Staying true to this unique little prototype will render success in any and all categories, for its engineering feat is totally connected, totally educational, and totally complete, thus perfectly organized. The following categories are laid out and transposed into an all in one and one in all formatted equation for the purpose of example and to expose its hidden order.

The Human Cell

Each and every human cell has a varied function, makeup, and position to fulfill, providing the many different traits that come together in order to create one unique, individual person. Even more awesome, all the features for that entire body, from the color of eyes and hair to the size of the feet, are genetically coded in each cell's DNA. Therefore, all cells in a human body, collectively and diversely, make up one overall design: a whole person, a single entity, unlike any other. And, on the flip side of that equation, the blueprint (one) for that complete individual can be found in every (all) individually different and unique cell. In the first half, the dominant emphasis is on the whole: how all cells harmoniously unite in a connected arrangement to create one body and one personality. In the second half, that whole is reduced to a skeletal abstract or a blueprint, recessively hidden inside each specialized cell, switching the main focus to the elaboration and importance of each and every individual cell. Every single cell has a unique and specific agenda all its own, yet, at the same time, it fulfills its collaborated role in the overall continuum. This phenomenon is described well in the following *Reader's Digest's* "ABC's of the Human Body" (pp. 22, 38):

> The body as a whole has been described as a community of cells, a social order in which each of 75 trillion individuals has some assigned place to occupy, some specific role to play....At the heart of every cell is its control center, called the nucleus. Inside the nucleus are 46 threadlike structures known as chromosomes, and contained in each chromosome are thousands of genes. Chromosomes and genes are made of deoxyribonucleic acid, or DNA, which transfers the hereditary material from generation to generation. The DNA of which every gene is composed contains a genetic message, a blueprint....As cells wear out, the body renews itself by a process of cell division called mitosis....It takes 30 minutes for the parent cell to split and become two daughter cells, each one an exact replica of the parent cell.

The Solar System

An encyclopedia designed for children describes the atom in a very candid manner: "In some ways an atom resembles a tiny solar system. At the center of the atom is a nucleus. Tiny particles of matter called electrons spin around the nucleus, somewhat as the planets spin around the sun." An old elementary science textbook simply defines the atom as "the smallest whole bit of each kind of matter." Likewise, the same could be said for the solar system: it is the smallest whole bit of each planetary system throughout our vast universe.

All of the elements and forces that go into the makeup of our solar system are so perfectly arranged and balanced that mankind is awestruck that time can be measured so precisely, in that seasons, eclipses, comets, etc. can be accurately predicted to the day. Yet, that same unique design, precision, and accuracy are modeled and uniformly present in the smallest whole bit of matter, the atom. It still may be hard for us to predict the daily weather, but, in contrast, time and long-range conditions can be accurately pinpointed in these two opposite bookends.

In conclusion, all the heavenly bodies in our large solar system, including earth, are arranged in this one gargantuan orbital design, and the pattern of that same unique design (one) is microscopically replicated in the whole kaleidoscope of matter (all) within that solar system, creating infinitesimal solar systems in and around us, tiny atoms. Paradoxically, when you delve into the smallest parts of our world, you can see the universe, and, when you observe the great wonders of the cosmos, you can see the familiar workings of our most intricate and miniscule parts. Therefore, these two opposites, which are basically the same thing manifested in extremely reverse poles, represent a continuum that spans a huge spectrum between the smallest whole bit on earth and the smallest whole bit in space.

The Color Spectrum

Even centuries earlier, symmetry in natural design was discovered and documented by some of our founding fathers in the field of science. For example, quoting from the book, *Astronomy the Cosmic Journey*, by William K. Hartmann, (p.121) he states: "An arrangement of all colors, in order of wavelength, is called the spectrum. Newton discovered he could see the spectrum of visible light by passing sunlight through a glass prism....Water droplets in a rainstorm act like little prisms and allow us to see spectrum—the same arrangement of colors from violet to red—in the rainbow." Thus, the organizing system that exists for the large is the same as it for the small, blocking in this continuum's bookends.

Consequently, the enormous amount of light coming down from the sun contains all possible colors bound together in one uniformly designed light spectrum, visible during rainstorms when we see the big violet to red rainbow in the sky. And that same color spectrum (one), in the exact same order, can be seen in all beams of light generated here on earth when split by a prism, creating an innumerable possibility of little rainbows.

From another perspective, the manner in which we are able to view these spectacular colors in our daily lives is made possible through the same consistent manifestations of the all-encompassing continuum. As the color white is the result of the whole spectrum of colors being reflected from an object to our eyesight, the reverse procedure is just as true. All colors absorbed into the pigmentation of an object reflect zero light rays to our eyes, resulting in the color we see or, should it be said, don't see as black. Therefore, the two bookends of the color system we experience are black and white, being the exact same process except in reverse, and the seemingly endless array of varied colors is the result of the pluses and minuses of these two poles. For example, when all colors but red are absorbed in an object and it is reflected, we see red. Accordingly, the same reflecting/absorbing order that applies to the bookends (all) can be seen in each and every (one) individual color as well.

Magnetism

Magnetism, perhaps more than anything else, best represents the dual manifestation and envelopment of repetitive chain reactions, such as the two mirrors facing each other mentioned earlier. This is one of the most consistent energy patterns scientists encounter throughout nature: multimagnetic fields within a larger magnetic field, such as the sun's plasma, and the earth's wind and weather patterns. Yet, scientists fail to connect magnetism to a definitive generalization. However, if you turn to children's textbooks and encyclopedias, they openly state the simple facts, for instance, "Magnetism is the force around a magnet….Did you know that the earth is a magnet? The earth has a north magnetic pole and a south magnetic pole….A magnetic field surrounds the earth. The magnetic lines of force around a current-carrying coil of wire look just like those around a bar magnet."

The huge magnetic field that surrounds the earth is exactly the same pattern as the magnetic field that surrounds any simple little magnet. Therefore, all things in our world are absorbed and taken over by earth's powerful motions: its pulls and pushes, lining up in one unique pattern—a giant magnetic field. And that one unique magnetic pattern can be seen in all earth's countless little magnetic fields, which are obviously infinite. Even the atom generates its own magnetic field. The following quote from the *World Book* CD-ROM, under "magnetic field," proves that everything produces a magnetic field since everything is made up of atoms: "The relationship between magnetism and electricity also operates in the atom. The motion of negatively charged electrons around a nucleus makes an electric current, which produces a magnetic field….In addition to circling the nucleus, an electron spins on its axis like a top. This motion also produces an electric current and a magnetic field."

Electromagnetism

A good educational source of current and easy-to-understand information is the classroom-designed "Assignment Discovery" program on television on the Discovery Channel. The lesson on magnetism and electromagnetism was so germane to this study that I copied the following narration verbatim:

> Electromagnetic fields that vibrate at right angles to each other create the waves that transport radio and television signals through space. Even light travels in the form of electromagnetic waves. Light is a vibration where the change in the electrical part induces the magnetic part, which induces the electrical part, which induces the magnetic part and the whole thing keeps inventing itself moving outward in a line, and that's what is coming out from the sun. The unified force of electromagnetism that makes a motor run also transports the light and energy that sustains life on earth.
>
> At MIT they are using magnets to create energy from fusion. The sun is nuclear fusion. Energy from the sun can be captured in magnetism here on earth and reproduce through this process. The hope to get energy from the ocean would be the nuclear fusion....They are discovering magnets are not just one of man's tools but electromagnetic force, itself, is key to unraveling the mysteries of creation. Einstein spent 30 years of his life trying to connect gravity, the primal force which keeps the earth in orbit around the sun and makes apples drop down off trees, with electromagnetism and try to combine them and unify them without success. But the unification objective is still with us. From the compass to creation, this mysterious force lures us into a world of discovery.

Thus, the electromagnetic patterns from our large sun are the same as the small atoms we are made of. The *Science and Invention Encyclopedia* (p. 190) states, "All atoms in their natural state emit and absorb pulses, or quantum, of energy, owing to the switching back and forth of electrons from one orbit, or energy level, around the central nucleus to another." Consequently, the whole diverse spectrum of energy (all) produced by the enormous sun manifests itself in one law of motion: electromagnetism. And electromagnetism itself (one) uniquely stands alone as the basic, self-sustaining unit of energy in all the multi-faceted transactions and motions of our little world.

Electromagnetic Radiation

Electromagnetism emitted from the sun to the earth also creates another extremely fundamental all in one continuum. Quoting again from the *Science and Invention Encyclopedia* (pp. 836–838):

> Gamma rays, X rays, ultraviolet radiation, visible light, infrared (heat radiation) and wireless (radio) waves are all of the same nature and can all be expressed in terms of a continuous interchange of magnetic and electric energy, each of which pulsates in a plane at right angles to the direction of travel of the radiant waves.

The following information about the universal behavior of radiating energy is another excerpt from *Astronomy, the Cosmic Journey* by William K. Hartmann, (pp. 122, 123):

> All light, of whatever wavelength—gamma ray, X ray, ultraviolet, visible, infrared, microwave, or radio—is electromagnetic radiation....In fact, this type of radiation is emitted all the time, all around us. Electrons (and atoms and molecules) in gases, liquids, and solid objects are in constant motion, jostling each other....Because the electrons of any object are being constantly disturbed by thermal (heat) motions, all objects continuously radiate a continuum spectrum of electromagnetic radiation.
>
> Ordinarily a nail emits no visible radiation....Despite appearances, the nail is radiating regardless of its temperature. When it is at room temperature, the light it radiates is so red (of such long wavelength) that we cannot see it because our eyes are not sensitive to infrared radiation...If the nail could be heated enough, its radiation would become distinctly bluish....A hot enough object radiates ultraviolet light, whose wavelength is too short for us to see.

Therefore, in conclusion, all energy from the sun comes to us in one special package, an electromagnetic radiation spectrum that provides us with the benefits of heat, energy, visible light, an information highway, and many medical/technical advances, etc., each in its own wavelength order. And that same unique pattern (one) of energy is generated or reactivated in all the relatively smaller objects or parts here on earth with the same wavelength spectrum and standard of measurement, very effectively representing the whole.

Consequently, if all energy from the sun comes to us in a continuum of radiation (one) and all objects on earth radiate or reflect a portion of this energy themselves in a similar mini-continuum of their own, then a spectrum (one) of the exact same order is reflected in all energies. As a result of this demonstrated, established order, along with forthcoming convincing examples, this equation could also be transferred to another level of understanding: all things in nature function in the arrangement of a continuum (one) and a continuum (one) is in all things.

The Basic Force of Energy

The following excerpts are taken from various elementary science textbooks, collectively, describing the same fundamentals of basic nonpoliticized scientific laws:

> A force is something that pulls or pushes. Work is motion against resistance. Just as there are magnetic lines of force for magnets, so there are electrical lines of force for electric charges. Electric lines of force have the power to push or pull on electric charges. Magnetic lines of force push or pull on magnets. There are two kinds of electric charges—positive and negative. Like charges repel and unlike charges attract.
>
> The molecules that make up the world of matter are bound to each other by electrical charges. The stronger the charges binding molecules together, the harder the material is. But the molecules never stop moving. Even in a block of iron, molecules vibrate back and forth. The same law applies to atoms: Atoms have, at their centers, tiny nuclei made up of even tinier subatomic particles. The attractions that hold the subatomic particles together are even stronger than those that hold atoms in place.
>
> Strange as it seems, electric motors run by a series of magnetic pushes. One common manifestation of electromagnetism is that a current flowing in a wire produces a magnetic field—this is the operating principle of an electromagnet, and can be harnessed to produce motion in electric motors through the attractive and repulsive forces of magnet fields. When a magnet is moved near an electrical conductor, turbulent eddy currents are induced in the conductor and it experiences a 'dragging force.' This dragging force can be used to produce motion, and conversely, the eddy currents (such as alternators and dynamos). This is an example of a moving magnetic field producing an electric current.

Perhaps Einstein's failure to prove his longtime suspicions that gravity and electromagnetism were one and the same was mainly due to complexity instead of simplicity. Basic answers are found in simple basic patterns first, not in complex mathematical equations. They have to be determined at the ground-floor level, not on the top floor. To our shame, the elusive answer to what gravity is is elementary to the question: what is a pull? A pull is simply the opposite of a push and vice versa. This is something a child could answer. Each one is half a process, and both sides are totally dependent on each other. You can't have one without the other; they are the same thing except in reverse. This pair of opposites makes up the fundamental core, the source, and the unifying principle of all forces: the duality transaction that is continually squashed and denied.

If all matter from the tiny atom to the giant universe were not subject to an equal force of repulsion (a push), then what would stop fusion (gravity) from attracting everything towards one giant explosion? On the other hand, attraction is what stops the repulsion of fission from expanding beyond its controlled destination. One year, the astronomy experts had us fearful that the universe was going to continue expanding from its core of origination, in which all of mankind would freeze to death sometime in the future. Then just a few years later, they reversed that theory and said the universe is most likely eventually going to collapse itself into one giant black hole. Instead, the incredible balance, stability, and motion we experience in our world is not just due to the force of gravity, which science books define as the force of pull (attraction) but which is also due to the equal counterpart of its magnetic field, the force of push (repulsion). Whether it is a question of electric motors, spinning electrons around a nucleus, planets orbiting the sun, or galaxies circling a black hole, they are all caused by the only self-existing, self-feeding, balancing energy we know: electromagnetism (pushes and pulls).

Therefore, the consistent pattern stands: all subjugated forces of energy line up and function within an overall giant push and pull electromagnetic field (one). And that same push and pull pattern (one) is found as the primal force in each and every (all) varied form of energy within that field, no matter how small. From the outer reaches of the universe to the subatomic particles that fashion our world, through electrical and magnetic charges, all energies thereof function together in one simple but most basic, bottom-line pattern: pushes and pulls.

To sum it up, from the stars and planets to earth's ecosystem, all experience the same pattern of opposites: from birth to death, from newborns to corpses, from new cells to dead cells, from destruction to renewal, from consumption to expulsion, from nights to days, from sleeping to waking, from springs to falls, from summers to winters, from volcanoes to earthquakes, and from black holes to supernovas. Thus, all things, from a rock to a planet, from an atom to the solar system, and from the heart to the lungs, each and everything in and of this universe is a living, breathing (in and out/push and pull) organism because of one basic, bottom-line pattern: the pushes and pulls of electromagnetism.

The Cat Family

There is one unique animal known as the "cat" throughout the entire animal kingdom. Accordingly, all cats, from the least to the greatest, the tame domestic house pet to the wild king of the beasts, collectively, fill a vast array of variations within the cat family (one), forming a characteristic outline, both behavioral and physical, unlike any other animal. And that basic profile (one) makes every cat (all), among all the extreme differences and tremendous variations in that species as well as individual uniqueness, undeniably a "cat."

The classification of the cat family is, of course, just an example of all species in the animal kingdom. With such a multitude of species, varieties, and types, it is easy to think there are no rules or boundaries. Even with such a wide selection of evolutionary differences, breeding combinations, and mutations, it nevertheless ends with the categorical continuum of its bookends. Without those two basic poles, there could be no true classifications in science.

The Human Thumbprint

Collectively and diversely, all human thumbprints that have ever been or will ever be, fit into the same category and compendium (one) that identifies them as being the composite image of a human thumb. Each and every individual thumbprint, fitting the basic profile (one) of a human thumb, is uniquely different from all other thumbprints; no two are exactly alike or ever will be, yet it never manifests characteristics beyond the boundaries of being a human thumb.

Of course, the same case could be made about any human characteristic or any category for that matter. Naturally, therefore, the above framework or template would also apply to human beings. Scientists have found that one's brain is as individual as one's face. Everyone's brain has the same physical features, but no one brain looks exactly like any other brain.

The world has its own method for classifying people. However, regardless of how societies measure a person's life or worth, according to their own standard of importance or worthiness, this unfolding blueprint of creation reveals the uniqueness of each and every person, no matter where he or she falls in the spectrum of humanity. Each fulfills a necessary and unique spot unlike any other, yet we are all basically the same.

A Government

Switching gears from natural design and causes to the inventions of man, what makes a perfect government? The ingredients that make a government successful are not estranged from all other categories that exist in nature or come from mankind. As shown earlier, the success of anything requires an equal balance of opposites. Therefore, the more categorical balances there are in a government, the more successful it is. For instance, the better the balance between the powers—of the people and for the people, of genders, of military and mind (body and soul), of domestic and foreign affairs, of physical and spiritual matters, and of federal and state—the greater is the government. The checks and balances within these duo divisions create overall stability and growth, with each dimension adding to the overall success.

Each one of these social opposites can be set up in a continuum of bookends all its own, for example, federal and state. The success of any political institution depends upon its match up to the all in one organization, such as the following: all the states in the United States come together in unity under one whole governing body, the federal government, forming one large organized, democratic structure. And that same structure (one) is carried out in each of its smaller governing states (all), like mini-federal governments being mimicked in each one. Yet, each state stands alone in its own diversification and unique representation of the whole (federal) in a way that none of the others can. Beyond the basic criteria making it a state, no two are exactly alike.

Going even further down the ladder into smaller governing bodies, such as city government, county seats, and precincts, the same prototype should be found. Each county seat has its own little capitol building, a courthouse, being the nucleus or axis for its jurisdiction, fashioned in the same manner as the nation's Capitol Building in Washington DC. In fact, the better the representation of the whole as well as the greater the diversity within those specifications, the better are the governing institution and the political arena, providing less dissention and corruption.

A Society

A society is the description given to a group of people in accordance to how they live together in neighborhoods, communities, states, nations, or a world of nations. The dictionary defines society as "a group of people having geographical boundaries and sharing certain characteristics, as language, culture, etc." Therefore, societies are primarily made up of families in a certain location that share the circumstances of their history, the conditions of their area, population, language, education, religion, resources, cultures, loyalties, and basic interests in varying degrees. Even though all societies are basically made up of all the above, no two are exactly alike.

The same can be said for the family unit: What makes it similar also makes it different. A family is not only the result of its own individual situation, making it uniquely different from any other family, but it is also a by-product of its own society, with all its own particulars. All families across the globe have the same basic needs, wants, and structure, but each family will not only be modeled after its own unique circumstances but will take on the characteristics of its own society as well. The families determine the society, and the collective characteristics of that society are in turn reflected in each of its families.

In conclusion, a French family will not only exude its own uniqueness as a family unit within the world of families, but it will also be a reflection of its French society, whereas a German family will reflect its German society, and so forth. Thus, all families that collectively and diversely come together, (in) under one country, form one unique society unlike any other society. And the basic characteristics of that society (one) are manifested in every family (all), which, at the same time, stands alone in its own individually unique reflection unlike any other.

A Filing System

What is the perfect filing system? In organizing files, the first initial task is gathering all material to be filed together in one place. That place is usually a file cabinet or a set of file cabinets. The next procedure would be to sort the material alphabetically into large file folders, marked A to Z, from beginning to end (from the least to the greatest) and to arrange them as such in the cabinet or set of cabinets.

Then, in order to complete and perfect the filing system, you would insert a series of smaller folders (such as manila folders), also labeled A to Z, in each and every file folder of the main alphabetical set. Consequently, all files would be organized in one large alphabetized arrangement to contain and create a systemized foundation for storing and retrieving information. And the same organizing alphabetical pattern (one) used in the overall set is used again in each and every (all) small folder for its own uniquely individualized system; no two are exactly alike, yet each is a part of and represents the whole (the larger alphabet).

A Good Story

A good story is broken up and composed of many little connecting stories, and, when the plot is arranged in a fundamentally sound format, it makes a great book. Each of the individual stories should be separate, distinct, and complete within itself, but connected by the same leading idea. In fact, the more extremely diverse the events and circumstances within the set story line, the more depth it has. The more twists and turns there are, the thicker the plot becomes. However, being structured within a proper paradoxical continuum, the opposing extremes will assuredly be checked and balanced.

Therefore, in a good story all the little stories and events will collectively and diversely unite into one big continuously flowing story, and that unique arrangement, how it all systematically comes together (all in one), is what forms the basic story line for the second half of the organizing continuum (one in all). This abstracted summarization of the overall story then becomes the outline, theme, or leading idea (one) that should be repeated and reflected in all the varied parts or chapters, but each stands alone in its own contrasting uniqueness, adding a different variational slant, circumstance, and point in time that none of the others can.

In retrospect, every part or chapter is uniquely different but reflective of the whole, and the whole envelops, anchors, methodically guides, and outlines all its many varied, contrasting parts. When you see the little picture, you should see the big picture, and, when you see the big picture, you should see the little picture as well, providing the reader with fascinating contrasts, structure, guidance, continuity, and connectivity, thereby making it a great piece of work.

Principally, what you have in all the above categories are classified equations based on one, equally divided into its two equal opposites. The first half of the continuum is the exact opposite of the second half, yet they are basically the same thing except in reverse, hence making each one of these exemplary, all in one and one in all continuums a paradox: two sides that are opposite yet complimentary to each other. Each needs the other side to make it a completely whole unit: $\frac{1}{2} + \frac{1}{2} = 1$. For example, having a place for everything is nice, but that is only half the goal; having everything in its place is the ideal and perfect scenario. Furthermore, the two halves are intrinsically linked together, because having a designated place is what makes it possible for the second half to do its job. And, vice versa, when the second half completes its duties properly, the whole operation has come full circle in a perfectly organized system. How well-organized the first collaborating arrangement is determines the success of the latter also. Then, in reverse, the more that is known and distinct in every thing, the easier it is to decide properly each one's important slot. For efficiency to prevail, each is dependent on the other for its own functions and fulfillments.

The organizing features of the all in one and one in all structure do not name a category but state, in the terms of the structure, the exact procedure as to how a classification should be organized. This demonstrates that, whatever is selected as a category for examination and classification, a successful and complete study or arrangement is guaranteed if it has been set up according to this scientific, mathematical template. No other design offers the ultimate balanced viewpoint as this all in one package does: absolutism and infinity both at the same time. True to its law, no matter what the category, there cannot be one opposite without the other. The ingenious composition of this continuum houses both opposites. The deductive elimination process associated with a contradiction does not apply to these paradoxical contrasts. The fixed, absolute bookends and the infinite proliferation generated from these two oppositions, creating a wide middle range of possible variations, are far from being contradictions. There is a need and a place for both.

Having pinpointed and examined the formula for perfect organization in this chapter, the next step is to go within the paradoxical format itself to find the next level of understanding. And, fortunately for us, the highly sophisticated, intricately designed paradox does more than just organize: it also teaches. The following probe into the virtues of this incredible dichotomy discloses, step by step, its unique arrangement and reveals its magical secrets by laying the truth virtually on the line.

4

The Educational Springboard

The initial procedure that helped tremendously in understanding the simple but complex anatomy of life's organized continuums, revealed in the last chapter, is the exact same method that will enlighten us in each category we encounter in our everyday lives, which is breaking one thing down into two's. As mentioned before, the beginning stages of awareness and comprehension begin with the basic, bottom-line, definitive contrasts of two opposites. Ironically, in all support systems, whether environmental or personal, it is all too easy for what's most supportive in our lives to go unnoticed completely. Why would it not be the same for nature's hidden agenda or the basic format of life, especially if we are not looking for opposites in the first place? In fact, oppositions and paradoxes are usually what people try to avoid. Consequently, it shouldn't be surprising that life's foundational building blocks, which are based on opposites, would also be something to which we are commonly oblivious. This naturally explains why the prototype for sustaining and reproducing life would be ubiquitously right under everyone's noses, but not so easily found; thus, the greater the support system, the greater is the obliviousness to it.

Therefore, in the beginning, the quest for true knowledge should start out simple, unassuming, and basic, not complex, conjectured, or sophisticated. Besides the single unit of one on which all things are based, you cannot get any simpler or more basic than opposites. A simple equation of equally divided opposites instantly and automatically provides the exact, definitive identity for each other. Why would it take a fifty-page essay to define a woman, for instance, when you already know what a man is? Or why would you go to great lengths to describe to someone how to identify hot when all you need to do is have that person feel something cold? Furthermore, when the cross-examining process initially begins, it must be stripped down to its most basic, contrasting elements or characteristics. Then, as you cross-reference the two opposites in the continuum, the learning process will begin to expand in direct proportion to each other: the more you learn of one side, the more you will learn of the other equally.

In summary, if you are not setting up your learning curve by first looking for opposites (the two bookends), the beginning process has not begun. The awareness process is still in the dark even if you are on the forty-ninth page of the essay. Unfortunately, it is rather difficult to follow these elementary procedures when the educational society still frowns on the duality concept yet, ironically, practices it when categorizing classifications. However, a real scientific search into this matter could lay this controversy on the line and finally resolve it once and for all. Then, what is now considered contradictory could be the interlocking key that will eventually open the door to all categories and all knowledge.

Laws in Action

As the old saying goes, "the proof of the pudding is in the eating." Nothing settles an issue like putting words and laws into action. Words are tremendous tools in education, but, to activate and complete the learning process, they must be acted out. Here again, one opposite needs the other to be effective or whole. Words without action are just as incomplete or ineffective as action without words. For example, if you are well versed in the essential laws toward understanding the paradox, as stated in this book, without seeing how they apply in real-life experiences, the concepts will remain vague and aloof. Equally inept is observing nature's active transactions without perceiving any structure, associations, or analytical connections.

One of the easiest demonstrations of how the laws of opposites work is something everyone is familiar with on a daily basis—our hands. This and the rest of the topics below are studied in a little different light than the all in one and one in all continuums, which dealt primarily with proper structure and organization, whereas the following equations deal more with the transactions and relationship of the two main opposite components in a given category. If you observe the simple interactions of these well-known subjects and how they relate to paradoxical laws within a unified continuum, their subtle and underlying structure can easily be revealed.

The Human Hands

The human hands are an incredible piece of machinery. No one knows this better than those misfortunate ones that have lost one or both in an accident. The hands perform so well and so often that their need can easily go unnoticed. Not only do we use them to grasp things (pull) or to punch with (push), but they are used a lot in our communications to others as well as communications to ourselves, in the form of sensory feelings. Notwithstanding, they create the same unique pattern that is in all things: two basic opposites.

How our hands work together in paradoxical fashion are the same in this unit as in any other unit. There is a right hand and a left hand, exactly alike yet completely divided and in opposition. They work in unison but in reverse form, fitting together like "hand in glove." Not only are their directions opposite, but the finger positionings are opposite to each other, each being an exact carbon copy and flip-flop of the other. Furthermore, each hand is complete within itself, containing everything the other has, and it can function alone on a limited basis. Each can singularly perform the same duty as the other; however, each needs the other in reverse cooperation to form a larger unit for tasks that are much more sophisticated and powerful.

However, the most unique characteristic forming the laws of opposites is the identity feature. Opposites are our main and basic learning tool. Putting that law into action, consider that, if all human beings had only one hand, there would be no such thing as a left hand and a right hand. This dual division is what creates each other's differentiating identity. Then taking that scenario a little farther out, imagine that we were endowed with two of everything except for the hands. Subsequently, if we understood the laws of opposites, we would automatically know that the missing hand would have to be the exact flip-flop of the one we already have, provided we were in the business of creating. One opposite is the guiding light or hand for the other and vice versa.

The Heart

The heart is divided into left and right sides that pump at the same time. Each is a pump doing the same thing, only in reverse, and with opposite substances. Veins collect blood from throughout the body and carry it to the right-side pump. That pump then sends blood to the lungs, where it picks up oxygen. Then, the left-side pump does the exact opposite procedure: the oxygenated blood flows to it from the lungs, which in turn pumps it through arteries to the rest of the body. "The steady pumping of the heart supports life by moving blood through the body. As it flows, blood delivers food and oxygen to all the body's cells and carries away wastes. Blood returns to the heart carrying a waste gas called carbon dioxide that cells produce as they use oxygen to obtain energy from food."

(*World Book* CD-ROM) The heart's two equal and opposite parts work simultaneously with each other, being equally and vitally important.

"The beat of your heart is produced by a tissue with special properties in the right upper chamber, or atrium. This center acts as a natural electrical pacemaker. The pacemaker—a kind of spark plug—fires an electrical impulse that causes the muscle fibers of both upper chambers to contract. The contraction forces blood both forward into the ventricle and backward where it causes pulsation of vessels. Just milliseconds after the pacemaker fires, its electrical charge reaches a second piece of specialized tissue, which is made of slow conducting muscle cells. This center relays the charge after only a tenth of a second or so. The charge then excites the muscles of the ventricles, which then squeeze the blood in the heart, increasing its pressure. This is the force that closes the valves between atria and ventricles and opens the valves of the blood vessels going to the lungs (right ventricle) and rest of the body (left ventricle). When the contraction ends, the pressure is higher in the circulation to the lungs than in the right ventricle, and higher in the arteries than it is in the left ventricle. This closes the valves, preventing backflow. The heart's "lub dub" sounds are related to muscle and valve movement." (*ABC's of the Human Body*, p. 90)

Thus, besides the back and forth, push and pull movements, the heart features another great pair of opposites: electromagnetism. The electromagnetic field is mainly responsible for the rhythmic contractions and relaxations. "Both sides of the heart pump blood at the same time. As the right ventricle contracts and sends blood to the lungs, the left ventricle also contracts and squeezes blood out to the body. The heart's cycle of activity has two stages, systole and diastole. Systole (pronounced SIHS tuh lee) occurs when the ventricles contract. Diastole (pronounced dy AS tuh lee) is the stage when the ventricles relax and the atria contract. One complete cycle of contraction and relaxation—called a cardiac cycle—makes up one heartbeat." (*World Book* CD-ROM)

Therefore, the heart is basically a pump generated by electromagnetism, which fundamentally creates and requires pushes and pulls. No part or function of the heart performs an isolated act or feature. There is a whole host of opposite pairs working dependently together, doing the exact opposite of what the other one does. Can you imagine a heart pumping blood in but not pumping it out? If the blood was not pulled into the heart, it could not be pushed out; if the blood was not pushed out, it could not be pumped in again. How simple, but oh how effective!

The Lungs

"Human beings have two lungs—a left lung and a right lung—which fill up most of the chest cavity....To supply oxygen to the blood and remove carbon dioxide from it, the lungs need to draw in fresh gas and expel stale gas. Fresh gas is drawn in when the diaphragm and other muscles in the chest wall contract. This action—called inspiration or inhalation—makes the chest volume larger and causes the lungs to expand. The expansion lowers the pressure in the lungs, and air from the atmosphere flows in. When the muscles relax, the lungs return to a smaller volume, and gas flows out into the atmosphere. This action is called expiration or exhalation." (*World Book* CD-ROM) Each lung is basically doing the same thing as the other except in reverse positions.

Simply put, when you are inhaling, the diaphragm contracts and the ribs expand, allowing for the pulling in of air. During the reverse procedure, when you are exhaling, the diaphragm is expanded and the ribs contract so that gas is pushed out of the lungs. To eliminate either half of this equal and opposite transaction would abort the vital act of breathing. The way the lungs work is just another wonderful example of the consistent pattern of the basic, vibrating, back and forth motion seen in all things, from the atom to the universe.

On the other hand, the seemingly simple, involuntary process of breathing requires the complication of all parts and all processes working simultaneously in dual cooperation in exact opposite interactions. Each part plays a dominant and recessive role at different times in reverse form. Even though each works independently from within, it also is designed to work as half of a whole function from without. People are known to exist with just one lung, but they are extremely limited compared to having both lungs. The two lungs and their functions together, meeting all the laws of opposites, are just another example of the unique paradox. Thus, breathing, one of the most vital functions and support systems performed by our bodies, goes practically unnoticed when it is working well.

The Muscles

Not only do the heart and lungs obey the laws of opposites, so do the muscles in our bodies. There are two sets of muscles that control skeletal movements, such as the raising and the lowering of the forearm. The motion is caused by contraction and extension (tension and relaxation or a pull and a push). The fibers in the muscle have the power of shortening, which is a contraction. When this happens, a bone is pulled toward it. For example, the bulge you feel on the front of your upper arm when you bend your elbow is the biceps muscle. The triceps muscle lies on the backside of the upper arm, and, when this muscle contracts, it straightens the elbow and stretches the biceps. Here again, we see the spontaneous cooperation of equal and opposite reactions. If the triceps muscle did not relax and elongate (extend or push the fibers apart), the biceps muscle could not contract (shorten or pull together), allowing the elbow to bend. Nor could the arm straighten up if the biceps muscle did not relax and allow its opposite component to do its job.

Muscles and bones not only respond together in this fundamentally dual setup, but they also work in conjunction with another dynamic push and pull duo. The orders that are given to the muscles to contract or extend mechanically comes from the body's energy source: the opposite, electromagnetic exchanges of matter and energy in the form of chemicals and neurons. Not only do the muscles obey the rhythmic spontaneity of contractions and extensions, but their necessary component, nerve energy, also moves through the muscles in an interdependent exchange of opposites.

The Brain

The brain is split into two basic halves: the right hemisphere and the left hemisphere. The motor cortex of the left hemisphere controls movement on the right side of the body, and the motor cortex of the right hemisphere controls movement on the left side of the body. These two major divisions of the brain have the same basic procedures and duties only in reverse order. Each is completely divided, working in opposition to the other, yet each needs the other to complete the brain's balanced functions. Not only do these physical connections work in opposite directions, but so do the mental tasks they perform.

The right hemisphere is the main center for musical ability, the recognition of faces, complicated visual patterns, and the expression of emotion, whereas the left hemisphere is the main center for language, mathematics, and logic. "Brain lateralization—the specialization of each hemisphere for certain functions—is not an either-or proposition. Sometimes both halves of the brain can accomplish the same task, though each may do so under special circumstances." (*ABC's of the Human Body*, p. 55)

Therefore, according to the laws of opposites, each side must have its own dual set of dominant and recessive traits that match up to each other. This makes each half complete within itself, doing similar functions only in a reversed and opposite manner. What is dominant in one is recessive in the other and vice versa. Each needs the other side to balance its own proportions, working together across that line of division in a magnetic connection, creating great mental tasks together in perfect harmony.

The Human Body

No matter how sophisticated and mind-boggling the human body is, it can still be broken down into two main and opposite parts: mind and body. These divisions can be interpreted in other terms, such as thought and matter, information and medium, or mental and physical. Furthermore, the mind and the body are also subdivided into these same two opposites themselves. Both the mind and the body can be broken down into mental and physical components, because these two opposites are intrinsically engaged on both sides of the equation. The mind cannot function without the necessary support of the physical system, nor can the body perform its physical activities without the vital information needed from the command center.

The previous category dwelt only with the physical functions and aspect of the brain, but its opposite, thought, is a lot more difficult to trace because it is invisible. However, following the laws of opposites, these two opponents are to be found coexisting: one does not produce the other, yet they are bound interdependently together. Also consistent with that set of proven laws is the undeniable fact that they are the same thing except in reverse. Therefore, if we are to learn anything about this mysterious side of the brain, we must look to its opposite as our guide instead of creating our own set of laws as many do. And, concurrently, the more we learn of one, the more we will learn of the other equally.

We have examples all around us that physical objects do not create thought, yet the common consensus is that our physical brain matter and its functions produce our intangible thoughts. Radios, televisions, movies, tapes, CDs, DVDs, computers, etc. all reproduce thought and ideas, but they do not create it. It comes from real people, from afar: from real studios, for instance, that capture and transmit the originated thoughts. There are no natural laws existing that show that one opposite produces the other. Therefore, we should conclude that, according to all known patterns, our physical brains are no different than the technical devices used to receive, process, and store transmissions of prepackaged thought, but that they are interpreted in varying degrees, making one opposite a giver and the other a receiver. If we would only practice the natural laws in nature through the bifocal lenses of opposites, life would not be so confusing. The following quotation creates two confusions. First, it gives a slight and subtle impression that the physical processes in the brain create thought. Second comes the inevitable, proverbial wall that always blocks our ability to ever know how thinking occurs.

"Scientists have only an elementary understanding of the extraordinarily complicated processes of thinking and remembering. Thinking involves processing information over circuits in the association cortex and other parts of the brain. These circuits enable the brain to combine information stored in the memory with information gathered by the senses. Scientists are just beginning to understand the brain's simplest circuits. Forming abstract ideas and studying difficult subjects must require circuits of astonishing complexity. Some aspects of human thinking—such as religious or philosophical beliefs—are still beyond scientists' understanding and may always be." (*World Book* CD-ROM)

When we talk about our body, the dominant connotation is physical. However, the physical body also needs information and commands (thought) from its recessive nerve center in order to perform its own chemical and biological duties. On the other hand, when we talk about our mind, the dominant connotation is mental, but it too is in need of recessive physical transactions to carry out its instructional commands. There again, each opposite provides the function or information for the other, each working cooperatively and in conjunction with one another. In summary, the most assuring thing about this study is that the mysterious ways of nature, whether mind or matter, can be understood when using the right educational tools—opposites, letting one side, such as the tangible, the easiest to comprehend, be the guiding light for the other, the intangible, for they are the same thing except in reverse.

The Human Being

The human being basically comes in two forms—male and female. Each is a duplicate of the other except in reverse; what is dominant in one is recessive in the other and vice versa. The dominant traits are designed for the purpose to give, and the recessive traits to receive. For example, in the reproductive process, the dominant genitals of the male give new life, and the dominant breasts of the female give nourishment to that life. Each needs the other to create a larger, balanced, and reproductive unit, having opposite roles to play. Even though there are exceptions, such as some women being stronger than some men and some men being better at domestics than some women, overall, throughout all societies, men have dominantly been the military strength of a nation and women the heart and soul of a country's culture. Thus, collectively and transcendentally, one represents the mind (woman/love) and the other the body (man/war).

Each gender receives its own identity from the other. If there were no such thing as woman, there would be no such thing as man and vice versa, each being a backdrop for the other in the discerning process. The more you learn about masculinity, the more you will understand about femininity and vice versa. Imagine a hypothetical society where there were only males being visited by a man from a normal two-gender society. This latter person could talk until he was blue in the face explaining to these people all the characteristics that make them men, getting virtually nowhere. However, the moment he brings a woman onto the scene, she would not only be a revealing sight, but now they could easily understand that they were men.

Each fits the full criteria for a human being, complete within itself. Both genders have everything the other has, masculinity and femininity, male and female hormones, only in proportionally reversed levels. However, when these two opposites attract and match up (the dominant to the recessive) forming a greater unit, they produce something quite unique: offspring. Scientists have even noted that the vagina and ovaries are quite similar to *the penis and testes* but in opposite directions, one inward, the other outward. When a fetus in the womb is three months old, the sex is not distinct. Scientists believe that, from that age forward, the hormones determine whether the sex organs grow inward or outward, depending on the gender, thus becoming dominant or recessive/giver or receiver. Therefore, each gender's reproductive organs are the reversal of the other, making the two a perfect magnetic (positive and negative) match.

The Globe

If, hypothetically, our globe could be sliced halfway down the middle along the equator and the two halves set side by side, they would look quite similar. The climate and terrain from the North Pole to the equator is almost exactly like it is from the equator to the South Pole. The reason is that they are opposites, making them the same thing except in reverse. These are the earth's two basic divisions, and, even though they are in opposite positions, they work in conjunction with each other. Like the perfection peak or middle ground of a continuum, the halfway mark is the dividing line of the equator. This is where the two magnetic fields come together and meet, and this is also where they become opposites. The magnetic control center at the North Pole takes in all the Northern Hemisphere, and the magnetic control center at the South Pole takes in all the Southern Hemisphere.

All interactions that take place between these two bookends consistently follow the dictates of a repeating pattern. Whether it is the lines of the magnetic field that surround the earth or the seasonal weather and wind patterns, the same laws of motion that take place between the pushes and pulls of an electromagnetic system, the cooperative exchanges of hot and cold temperatures, or the equalization of weights and pressure apply to these two opposites as well. All the elements on earth, regardless of the medium (air, water, wind, etc.), follow the same paths of direction under the same set of rules. For example, air at the equator is warmed quickly and rises towards the poles. Then colder air from the north and south blows in towards the equator, replacing the warm air.

The same effect can be seen in water heated on a stove in a glass pot. When put over a flame, the bottom layers of water are warmed and start to rise. As the warm water rises, cooler water above it sinks to the bottom, forming a fast circular motion when heated to the boiling point. In a weather pattern, warm air expands and pushes upward and away from land. Aloft, the air flows out to sea where it cools and descends, pushing back again toward the land, creating the same circular motion found in all magnetic fields. The same recycling system can be seen in precipitation. As the warm air rises, water vapor is taken with it. When the air cools, the water vapor condenses and these tiny drops of water become clouds. As the drops get heavy enough, they will fall out of the clouds to the ground in the form of rain, snow, or sleet.

The following quotations from *The New Book of Knowledge Encyclopedia* could not explain this dual and reverse pattern any plainer or more factually:

> Air heated at the equator expands upward. It raises the pressure aloft and drives air toward the poles at upper levels of the atmosphere. At the same time, the air over the cold polar regions is contracting, which allows more air to pile on top of it. This raises the pressure near the ground. Air near the ground is driven toward the equator.
>
> If the earth and its atmosphere were not rotating, the wind would blow from high to low pressure. Because of the rotation of the earth, the large wind currents turn in such a way that high pressure is found on one side of a current and low pressure on the other. In the Northern Hemisphere, if you stand with your back to the wind—facing "downstream"—low pressure is found on your left and high pressure on your right. In the Southern Hemisphere, the reverse is true: the low pressure on your right, and the high pressure is on your left. This is known as Buys-Ballot's law of wind and pressure. (Christoph Buys-Ballot was a nineteenth-century Dutch weather scientist.)

The earth as a whole rotates from west to east. If you could stand off in space and look down at the North Pole, you would see that the Northern Hemisphere rotates counterclockwise. Looking up at the South Pole, you would see that the Southern Hemisphere rotates in a clockwise direction.

As inhabitants of the earth, we rotate with our planet. Therefore, whatever motions we see on the earth—such as the winds—are relative motions. We see them as if we were watching them from a rotating phonograph record. As the air comes toward us in the Northern Hemisphere, we move away from it in a counterclockwise direction. Therefore, if we are facing downstream it looks to us as if the air has turned to the right. In fact, it looks as if some force is pushing it to the right. This force is known as the Coriolis force, after a nineteenth-century French mathematician. It is earth's rotation. In the Southern Hemisphere the Coriolis force pushes to the left (when the observer is facing downstream).

As the air starts moving from high to low pressure in the Northern Hemisphere, the Coriolis force turns it to the right. But after a while the Coriolis force is balanced by the pressure force that always pushes from high to low pressure. Then there is no further turning of the wind, and it blows according to the law of Buys Ballot. Now you can begin to see how the easterly and westerly wind belts appear. As the air in the tropics begins to flow from the subtropical highs into the equatorial low, the Coriolis force turns it until it is blowing from east to west. Thus, the trade winds develop. As the air flows from latitude 30 degrees to latitude 60 degrees, the Coriolis force turns it until it is blowing from west to east. Thus, the westerlies develop. Similarly, the Coriolis force produces the polar easterlies.

One of the wind patterns found every day on weather maps is the cyclone. A cyclone is a region of low pressure surrounded by higher pressure. In the Northern Hemisphere the wind blows in a counterclockwise direction around the low-pressure center. In the Southern Hemisphere the wind blows in a clockwise direction. It is this circular motion around the low-pressure center that gives the cyclone its name....The opposite of a cyclone is an anticyclone. This is a region of high pressure surrounded by lower pressure. Around an anticyclone the wind blows clockwise in the Northern Hemisphere and counterclockwise in the Southern Hemisphere.

The bottom line, the reason why our globe works the way it does—its energy patterns, its weather patterns, its topography, its pushes and pulls, its equal and opposite responses, etc.—can all be explained in one simple equation. It is one whole unit divided into two equal parts that are the same thing, except in reverse. To know one is to know the other. Specifically, to know how a magnetic field works is to know how temperatures, pressures, winds, and weather work as well, and to know how all these transactions work together is to know simply how opposites work in an interdependent fashion, since they are basically and functionally one and the same.

The Environment

As far as the eyes can see and the ears can hear, even with the most sophisticated technical equipment available to mankind, it appears that there is only one habitat within our range of investigation that supports life. And, alas, this too is divided into two opposite life styles.

If we branch out into outer space and view our globe from a different angle but with the same bifocal lenses, a unique pattern will emerge that is so connected and awesome that it boggles the mind. When you study earth's atmosphere and its oceans, you will not only see another incredible ecosystem but another pair of opposites intrinsically linked together as well. The atmosphere above the ground is described as a sea of air, which is a mixture of gases, including water vapor, and which has weight (matter pulled by gravity). In this natural medium are travelers that must propel themselves with their wings. These birds and bugs can fly singly or in flocks. Closer to the ground where there is more oxygen, water, food, plants, and trees, lie their resting and nesting places. The dry land is the bottom of this spectrum, where life is teeming with earth-bound animals, plants, human beings, and dwelling places.

The exact opposite and flip-flop arrangement of this picture is what's below earth's surface, under the oceans. The terrain on the bottom of the sea is land much like the land above, with its own mountains, valleys, streams, plants, and aquatic creatures on the ground. Also, the ocean is a habitat just like the atmospheric one above it, with all the necessary ingredients to support life, such as food, water, oxygen, nitrogen, carbon dioxide, and, in lesser degrees, sunlight. And in this watery medium are travelers that propel themselves much the same way that birds fly through the air, moving their fins and/or tails almost in the same manner. In fact, if you look closely at how the rows of scales are orderly arranged on a fish, you can discover that they look much like the arrangement of feathers on a bird. Instead of flocks, fish group together in schools. Neither are places to stand but are highways of transportation: two mediums sharing the same dynamic laws of motion because their environments are basically the same thing except in reverse.

The way objects move through the air is basically the same as the way that objects move through the water, and the natural flows of the air and water themselves are virtually the same. "Aerodynamics (like hydrodynamics) is a branch of fluid dynamics, which is the study of fluids in motion. The fundamental laws governing the movements of gases, such as air, and liquids, such as water, are identical." (*Science & Invention Encyclopedia*, Vol.1, p. 26) In fact, studying the pressures and wave motions that created water travel helped tremendously in creating successful air travel for mankind. It has only been within the past two hundred years that human beings have cleverly released their bondage from land and are heavily infiltrating both air and sea, simply by studying the visible and simpler opposite first and then by following and transferring its basic patterns over to the more complex one, hence the true essence of learning.

And last, but not least, the recycling that takes place between these two opposites goes far beyond the obvious water cycle exchanges of rain and evaporation. It is one enormously complex ecosystem, the two being highly dependent on each other in many ways, thereby fulfilling the laws of a paradoxical continuum with dominant and recessive exchanges on both sides in order to support one another the way opposites do.

Our Audio and Visual World

Two very small segments of the energy spectrums we experience here on earth come in the form of senses. These two main avenues of information enable us to see and hear some of the activities taking place in and around our world. Light is a small portion of an enormous range of electromagnetic radiation that comes to us from the sun. The many different frequencies and wavelengths of this large continuum, which cannot be detected by the human visual system, lie on either side of the visible spectrum, such as infrared and ultraviolet. Sound is also a small segment of another large spectrum with different frequencies and wavelengths. The human ear cannot detect acoustic waves having frequencies of vibrations below twenty cycles per second, such as infrasonic, or above twenty thousand cycles per second, such as ultrasonic. Therefore, light and sound are two forms of energy that are basically the same thing except in reverse.

Their modes of travel follow the same basic patterns as well. For one, they both travel in vibrating waves. The vibrations of the electric and magnetic fields of light rays are at right angles to the direction the light is traveling. When a part of these rays enters the lenses of our eyes, they help create vision. Vibrations also help produce sound by creating patterns of compression and rarefaction in matter. As these molecules move back and forth, they collide with neighboring molecules. Thus, a series of wave disturbances are transmitted through air, liquids, or solids to our ears, causing our eardrums to vibrate to their frequency and create sound. Light waves travel fastest through air. The denser the matter, such as water, the slower is the travel. Sound waves are just the opposite. They travel fastest in solids and liquids, slower in air, which is the very reason why we see the lightning before we hear the thunder. I suppose, if lightning were to occur under water, we would hear the thunder first and see the flash of light second.

Light waves can be absorbed, reflected, or transmitted by objects. Most things you can see are visible because light from somewhere else bounces off them. Light bouncing off something is reflection. Sound waves can also be absorbed, reflected, or transmitted. Smooth, flat surfaces, such as walls, reflect sound waves. When you hear an echo, you hear the reflection of sound waves. Curtains and carpets absorb sound waves. "Being a wave motion of air molecules in the atmosphere, sound obeys the rules of reflection, diffraction, and dispersion in a similar fashion to the far shorter wavelength electromagnetic waves which we call light, but obviously from, and through, different materials." (*Science and Invention Encyclopedia*, Vol.1, p. 12)

Another factor that the two opposites share is refraction. "Since the temperature of a gas affects the velocity at which sound travels through it, atmospheric temperature variations can result in velocity changes and consequent direction changes of the sound. This change of direction is called refraction and is directly comparable to refraction in optics....Both light and sound waves can be diffracted, but because of different wavelengths the results are different. All waves normally travel in straight lines, but they can bend around obstacles whose size approximately equals the wavelength of the radiation. Because the sound waves are normally between several inches and feet in length, they bend around most commonly encountered objects, whereas the short wavelength of light means that ordinary objects do not diffract it. It is, therefore, possible to hear sounds from sources that cannot be seen by the observer." (*Science and Invention Encyclopedia*, Vol. 18, pp. 2467–2468)

Lightning and thunder are good examples of one incident that splits into two basic and opposite processes. Obviously, lightning comes from a high concentration of heat, being a part of the radiation spectrum that burns just about anything in its path; whereas sound waves deliver their punch without heat. Sound can sometimes hurt our ears, but it never burns them. Its bombardment has more to do with the collision of matter rather than electricity. Basically, one's hot and the other's not, which is the main reason ultrasound techniques are safe and less intrusive than ultralight. Therefore, there must be a separation at the moment of impact between these two forms of energy that we experience with our senses within split seconds of each other. Yet each has identical missions: providing knowledge of our world in two opposite forms, seeing and hearing. Each displays the same mechanical properties and patterns, forming a continuum or spectrum of its own when they are coming together, thereby exemplifying one of opposites' most unique laws: to be as much alike as they are different. These two opponents, like any pair of opposites, were designed to work harmoniously together, as modern technology is finding out.

"Humans cannot hear ultrasonic waves (sound), but the development of the acoustic microscope enables us to see with them....Using schlieren techniques and optical acoustics, complex sound patterns produced by reflecting a beam of sound off a structure can be photographed....Scanning laser acoustic microscopy (SLAM) uses sound vibrations to search out defects in metal parts at the micron level. Sound waves beamed through the parts are altered by defects and impinge on a sheet of gold foil, to produce distortions matching those of the defect, which are revealed by laser light." (*Science and Invention*, Vol. 18, pp. 2465–2466) In reverse, the opposite has been equally rewarding. "In September 1964 a note of 60,000 vibrations per second was achieved by striking a sapphire crystal with a laser beam (engineering acoustics). This is the highest musical note ever produced." (*Science and Invention*, Vol. 10, p. 1373) Without naming all the obvious electronic gadgets we have today, I can still say that using light (optics) to enhance its opposite component, sound, has produced a revolution in the world of entertainment: "ultralight" that we can hear.

Consequently, according to the fundamental laws of opposites, these two important features of nature, enabling us with the power to see and hear, work simultaneously with each other. One does not produce the other, and one does not eliminate the other; one's high or low intensity is simply matched by the other. We can also learn a great deal about them by applying another incredible law: side-by-side in a cross-reference examination of their inner parts and miraculous workings. The more we learn of one, the more we will learn of the other equally; and we have.

Principles Applied

These patterns should finally dispel the common and false notion that one opposite produces the other or that one empties into the other. All these opposites are found coexisting, being independently bound together in one unit. As has been proven and according to the laws of nature, matter and all its components cannot be destroyed; they just change their arrangement although the form they change into is not something strange, disconnected, or disorganized. They simply reverse their magnetic poles or switch to their opposite half on the other side of the equation. For, again, what is dominant on one side is recessive on the other side and vice versa. The simple reason is because they are the same thing except in reverse. They are definitely divided but, at the same time, united. That's what opposites do.

Therefore, matter can discharge energy because it contains energy, not because the matter is or turns into energy. The divided line remains: matter is matter and energy is energy. When matter is arranged in a subtle and relatively stable form, it becomes the dominant side, with energy being its recessive (hidden) trait. Just the opposite is true in reverse when it switches to the other half of the equation; then the dramatic change in form causes energy to be dominant and matter recessive (hidden). The same ingredients lie on both sides. What changes is the form or arrangement of all the parts, and it always lines up according to two poles: dominant and recessive. One does not turn into the other, but each contains and utilizes the other in a recessive form on its dominant side of the equational line. In other words, each opposite embraces its own opposite but in a recessive role.

Putting this whole chapter in summary and in perspective, the reason the productive or learning process basically comes in two's is because everything we come in contact with in our environment is set up in two's. All these personal, everyday subjects and experiences listed above, that we share in as human beings, prove that they are no different than the most basic, underlying, fundamental, building blocks of life, studied in previous chapters. Just as we learned earlier that, from their recessive to dominant roles, there is chaos (order we can't easily see) in order (order we do see) and order in chaos, there is energy in mass and mass in energy, there is basic in general and general in basic, and, above all, there is all in one and one in all. We can see this same process working in DNA and cells, atoms and solar systems, family and government, electricity and magnetism, light and sound, air and water, hot and cold, black and white, the northern and southern hemispheres of our globe, the animal kingdom, human beings, and the human body and all its parts.

All the examples above should be enough to demonstrate how consistently this pattern holds up in the educational process and that we can confidently rely on one side being the guiding light, example, impetus, or springboard for the other. However, there is one last stage involving the basic structural design of the paradox that will be addressed next. It is rather vague and flexible, which is in opposite fashion of the two fixed bookends, but it is just as real and just as vital.

5

The Quintessential Classroom

There is actually nothing of man or nature that I have found that cannot be learned, categorized, analyzed, measured, or monitored when set up in the platform of a continuum. As a matter of fact, real, clear-cut analysis of the truth, using the continuum guidelines, is not reserved for lessons taught in the classroom only, as a measuring tool, nor is it valid for just a very narrow margin of subjects. Ironically, as the list below indeed testifies to, the continuum is the quintessential classroom, no matter where you take it.

- It is the only way to organize and study a classification because of the essential act of gathering all things under one category for a full examination.

- It is the best way to analyze your subject of inquiry because the springboard for verification, after all the relative facts have been gathered, simply lies on the opposite half of the equation.

- It is the only truly helpful forum in the decision-making process for lining up pros and cons in order to view one's options clearly.

- The cross-referencing, cross-examining, definitive, contrasting opponents, represented by the two polar bookends, make it the perfect review board and learning curve.

- It is the most efficient method for measuring and monitoring success; the throttle to regulate balance and perfection simply rests in the midst, at the summit where the two halves meet, at a very clear, precise point, leaving out any guesswork and speculation.

- Last, but far from least, it is the only way to measure time and relativity accurately. Without a clear reference point to judge between differing and opposing forces, objects, or times, there can be no standard or criteria to draw from or measure.

Learning the true and perfect form of organization and mastering its structural feats will serve us well in our quest for knowledge as well as acquiring success in any and all of our pursuits whatever they be. Using the continuum as our prototype and making it applicable to our lives, we can follow its example in a step-by-step process only after we have thoroughly studied and understood its basic fundamental makeup.

The "Designing" Graph

Numeral one is the smallest, most basic, common denominator existing in all things. It is the foundation by which all things are built and the starting point at which we learn. All categories and all structures begin with one, and it is the platform upon which all things take place. Whether it is a person, place, thing, or action, all can be reduced to and represented as one unit. Likewise, a unit is what houses any classification be it math, family, society, science, or the elements. *The Doubleday Dictionary* (1975 edition) defines unity as "the state or quality of being one or united: oneness; The quality or fact of being a whole through the unification of separate or individual parts."

The following graph, illustrated by a linear line and labeled "phase one," symbolizes this one bottom line and single unit, being the first element, the beginning, and most basic part of the paradoxical design. All things begin and end here on this basic foundation:

PHASE ONE

The Basic Foundation

Once you have the basic substance upon which to build or the basic category upon which to gather information, the second step in the creative development process, no matter what it is, is separation: equal division of two opposites. Without this duality and polarization, we would not have awareness, definition, balance, precision, relativity, cross-referencing, productive transactions or, of course, magnetism.

This next graph is the second stage, phase two, which is where the reproductive as well as the awareness abilities begin. Also, a short course in biology would teach anyone that the growth and reproduction of any living thing, from a single cell to complicated organisms, is the result of a unit continuously dividing itself into two equal parts. Therefore, the second element and process of the continuum or paradoxical pattern is numeral two, which divides the single unit (one) into two equally opposite parts.

PHASE TWO

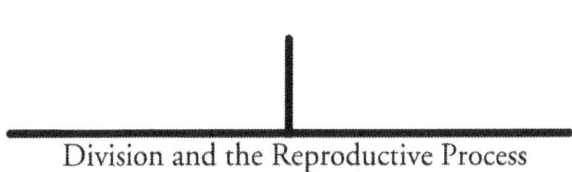

Division and the Reproductive Process

When two perfectly matched and divided opposites are bonded together in one unit, the third phase, numeral three, of the paradoxical design is produced that is more than the sum total of its two sides. This unification creates something beyond its polarized parts with a unique life of its own. It is what gives the spectrum the ironic image of being a seamless whole, and it is the element responsible for synthesizing its harmonic unity.

This is where the modern world of education is currently hung up. The objectionable experts cannot understand how the world can be divided and whole at the same time; therefore, most reject the unpopular paradoxical concept. However, the reality is that you cannot get to three until you first come to two, and you cannot become united until you first become divided. This simple and logical elementary math is applicable to all things. Phase two is fundamentally where the present world of education is stuck, being mainly divided into two schools of thought: those that believe the absurdity of unity or integration without clear-cut divisions, and those who believe you should have division without unity. Logic could possibly be on their side, in a small, distorted manner, if that was all there is to the equation. But if and when they ever get to phase three, they will see that neither is correct, and they will finally become aware that unity is the result of two equally divided opposites interacting in unison in a supporting position. On the other hand, they may come to realize how futile it is to try to keep two perfect opposites isolated from each other.

The pattern of two represents that which is fixed, precise, clear-cut, divided, and basic, such as two bookends, making it relatively easy to pinpoint. Whereas this third element or pattern represents all that is in between and beyond, making it extended, flexible, vague, and general but whole. As proven earlier, each of the two opposing sides is complete within itself, but, when they are joined together, something altogether different is produced that is, however, principally the same because it is a collective, generalized version of the two. Therefore, this third feature is itself an opposite to its dual interim components, making it as much like them as it is different according to the laws of opposites.

As stated earlier, in a well-organized unit, each individual part has to be able to stand alone in completeness and understanding, yet each must reveal and be a part of the whole. Not only must each half of an equation be complete, having all the same ingredients the other has in reverse proportions, but the same holds true for this third development as well. It too must represent the whole, and it does so, but in a different way than the two halves. Even though all the same ingredients are in each half, they are disproportionate, and each needs the other to make up for its deficit in order to make it an even match. For example, women have both male and female hormones, but her female hormones are dominant and her male hormones are recessive, whereas it is the precisely reverse arrangement in men. However, this third phase wraps around, takes in, and combines both sides with matching proportions, creating a larger whole.

The reproductive steps of creation are not complete until it gets to this extended third level, symbolized in the following graph as the third line that is drawn from one bookend to the other, crossing the line of division and unifying the two opposites. This consolidating feature is basically the same as the two equal sides plus more. It not only represents the combination, collective overview, and range of the two opposites coming together, thereby creating harmony, balance, and wholeness, but also creates something intangible and infinite, something far greater and beyond itself. In other words, as plainly as it could possibly be stated, a successful, single, categorical unit reproduces itself three times but in three different ways. Dissection of these basic fundamental steps of nature finally gives clarification and credibility to the phenomenon we often see happening (but didn't know why) in the common phrase "Everything comes in three's." The diagram below demonstrates this third phase with its three developmental steps:

THREE

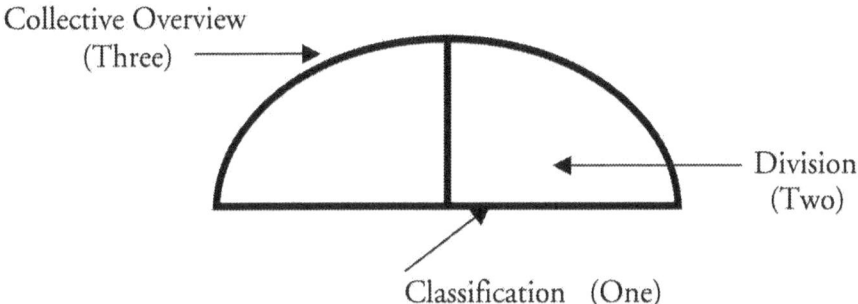

Collective Overview (Three)

Division (Two)

Classification (One)

A good example of this progressive process is the filing system. It would be quite an understatement to say that a good filing system is more than the sum total of its two basic bookends—a cabinet with folders prepared for files (a place for everything)—and more than the information stored in those files (everything in its place). Of course, what is much more important than a file cabinet full of information is what it helps to create beyond its drawers. From the steady business of recycling information (the storing and retrieving of files), great things can be and have been produced; thus, ultimately, when you match up another set of opposites, such as a great mind and a great filing system of information, great accomplishments can happen.

The same situation applies to anything organized in a place for everything and everything in its place continuum. It is not just the neat arrangement that is important here as it is the active results that come from a perfect recycling system. It is the many activities that take place in life and the accomplishments produced from such an orderly and efficient system that make it so much more than just the sum of its parts. Needless to say, anything and anyone could reap untold benefits if everything to be organized was set up with all things in their proper places, followed by the discipline of returning them to their exact spot when finished, a goal few of us achieve but how productive when applied.

One of the greatest examples of what exemplifies more than the sum total of its parts is the union of man and woman. The two genders generate more than romance and offspring. From this match up of opposites, reproductive effects are more far-reaching than just the result of sex, children, and families. For example, marriages (fathers and mothers) produce children (boys and girls/brothers and sisters), children produce families, families produce societies, societies produce countries, and countries produce governments, reflecting that very same two-gender order. Our sophisticated societies subtly reflect the same dual match up of femininity and masculinity as the simple union of a man and a woman. A country always splits itself between its domestic and foreign affairs, love and war, culture and military, arts and sports, tenderness and strength, mind and body, "beauty and beast," etc. One simply could not exist for long without the other.

The coming together of opposites creates unity, no matter what the category is. And the more balanced the two are, the more productive and harmonious is the union. For instance, history will prove that in every era or country inhibiting the full productivity of half this equation, which is its women, the society suffered or still suffers from backwardness, repression, and inhumanity. However, where women are as relatively free and equal as the men to exercise their half of the union, it will be a progressive and productive country, allowing for the full blossoming of its two contrasting sides. As in any mathematical equation, the figures do not add up if they are not equal. But when they are, each keeps the other in check as well as supports the other, bringing out the fullness and best in one another. Thus, harmony at the basic level can bring about harmony on many levels. The following diagram will demonstrate the varied, reverberating, and transcending results produced by this third, multiplying step of the creative development process when two equal, definitive opposites are joined together:

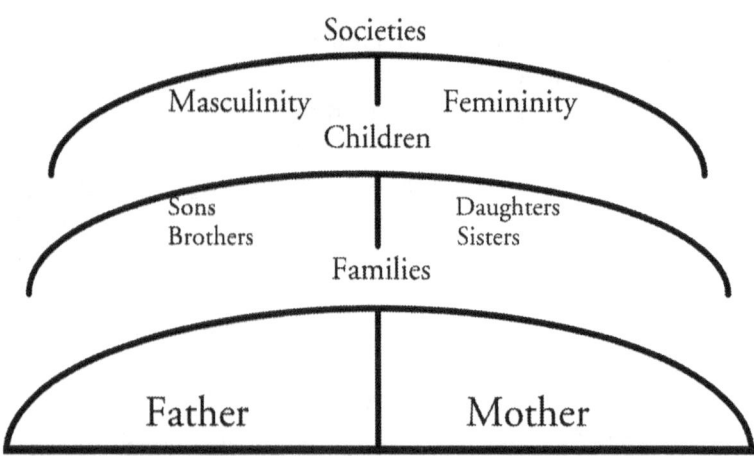

Life's Templates

Templates can be very helpful in many different areas. They are tried and true models that guide and simplify our work, set patterns that can be used consistently again and again. The following templates for organizing any and all classifications will demonstrate how life's basic fundamentals are already set up for us in these graphic and mathematical arrangements, giving us role models to emulate in our everyday lives, showing the universality of nature. The positive sign (+) represents the dominant and the negative sign (-) represents the recessive.

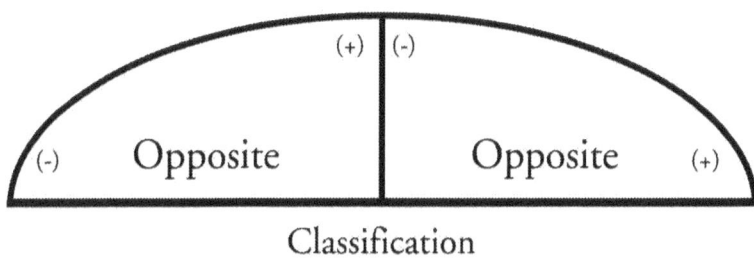

Classification

The template or pattern for classifying any category begins with one, giving it a single and basic label. Then the next step is to divide it into its two opposite bookends. The final step, three, is to lay out the reverse dominant and recessive components on both sides of the equation. What is dominant on one side will be recessive on the other side and vice versa, thus creating a whole by connecting and correlating the magnetic relationship that exists between the two opposing, divided halves.

"Putting the cart before the horse" can make it rather difficult in finding the two basic divisions, because so many times people will ask: "Well, if there's two of everything, then what's the opposite of such and such?" That is why a category needs to be named first, giving whatever that something is a label and a purpose,—a reason for being. This will make finding its polar ends much easier. Another exercise that is helpful is trying to see spectrums within a spectrum. As demonstrated, each side of the equation is a complete spectrum within itself yet is half of a larger unit at the same time.

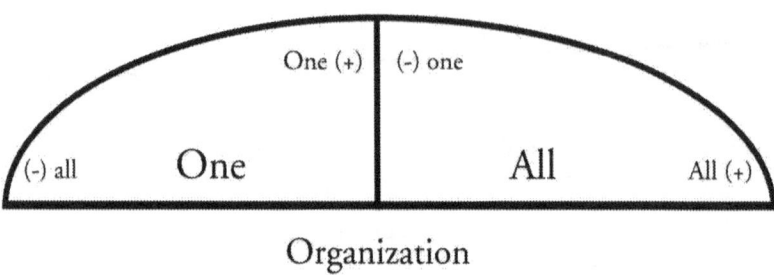

Organization

This graphic and its mathematical transactions is none other than the exquisite design of the perfect organizational continuum. Using this template for organizing anything will assure a masterful arrangement like no other, for it will include all things in their appropriate places, making it the only way to achieve the quintessential spectrum of a place for everything and everything in its place.

This basic all in one and one in all outline is set up in order to maximize thoroughness and provide complete representation, thereby setting the example to follow in all things and all categories, no matter what the subject. It is an organizing template or role model that can be used anywhere, thus making the best organization for one, the best organization for all, plus rendering one just as important as all and vice versa. Then, ultimately, in its final summation, as well as setting the standard for organizing all units, it can clearly be said that, when something is organized at its best, there is all in one and there is one in all.

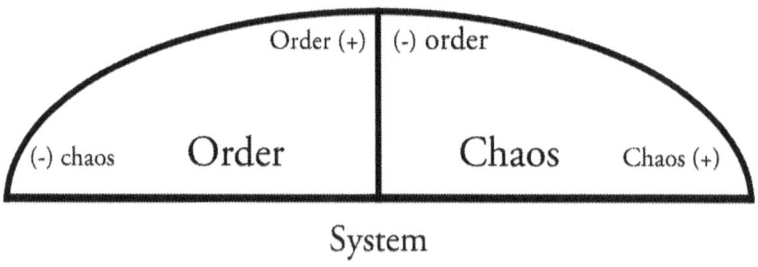

System

The definition of system is "a group or arrangement of parts, facts, phenomena, etc., that relate to or interact with each other in such a way as to form a whole." The definition of order is "established or existing state of things." Therefore, just like the classification and organization templates, we will find there are two opposite states, arrangements, or systems, as well: order and chaos.

Order is the same thing as chaos except in reverse. One is order we can recognize, and the other is difficult to comprehend but order nonetheless. Each embraces the hidden systematic order of the other. On one side of the equation, order is dominant and chaos recessive, and, on the other, chaos is dominant and order recessive. Thus, the same as the one and all template of organization, there is chaos in order and there is order in chaos. The following is a medley of revolutionary and astounding excerpts quoted from the book titled *Chaos*:

> Simple systems give rise to complex behavior. Complex systems give rise to simple behavior....The simplest systems are now seen to create extraordinarily difficult problems of predictability. Yet order arises spontaneously in those systems—chaos and order together....Only a new kind of science could begin to cross the great gulf between knowledge of what one thing does—one water molecule, one cell of heart tissue, one neuron—and what millions of them (all) do....Nature forms patterns. Some are orderly in space but disorderly in time, others orderly in time but disorderly in space.
>
> The heart of the new snowflake model is the essence of chaos: a delicate balance between forces of stability and forces of instability; a powerful interplay of forces on atomic scales and forces on everyday scales....In metals the molecular symmetry is different, and so are the characteristic crystals, which help determine an alloy's strength. But the mathematics is the same: the laws of pattern formation are universal.

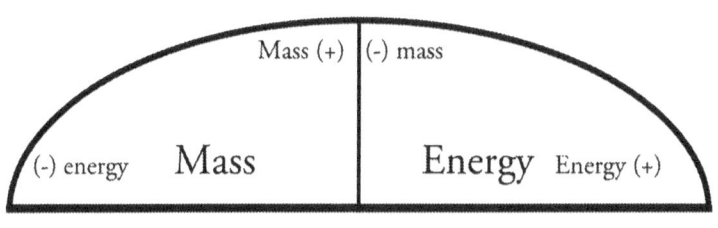

Physical Substance

The dictionary definition for substance is "the material of which anything is made or constituted...essential components or ideas." Therefore, any and all physical substances we as human beings experience can be arranged in basically two opposites: mass and energy. It can sometimes be hard to reduce a classification down to its two lowest components, and something that can be helpful is to build a pyramid allegorically towards finding the main point. Its broad base covers many areas, but, as you build blocks of information, the subject can be narrowed less and less until you reach the summit with a single item or idea that can't be reduced any further. Once you have the one category, then it can be divided into its two bookends, which will be basically the same thing except in reverse. The more you understand the mechanics and the purpose of your subject, the more you can get to the core or single issue of a category.

The relationship and laws that exist between mass and energy are the same as for order and chaos. They are, as Albert Einstein discovered, different forms of the same thing and interchangeable, which is exactly the same as saying that these two opposites are basically the same thing except in reverse. In one half of the equation, mass is dominant and energy recessive, whereas in the other half energy is dominant and mass recessive. Each dominant side envelops its opposite in a recessive fraction. In conclusion, the same format found in the previous templates, applies here as well because there is energy in mass and there is mass in energy.

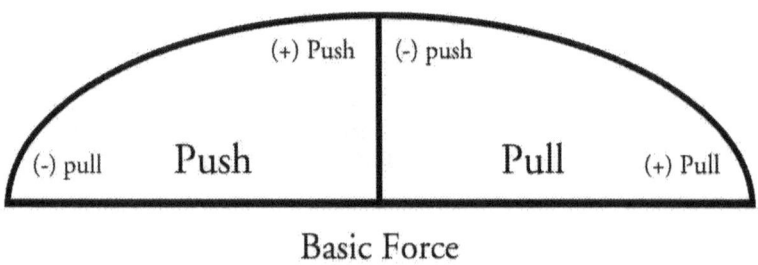

Basic Force

From the outer reaches of the universe to the subatomic particles that fashion our world, through electrical and magnetic charges, all energies thereof function together in one simple but most basic bottom-line pattern: pushes and pulls. A push is really just a pull in reverse and vice versa. They are opposites working together in cooperative unison; if one side did not recede, the other could not proceed.

Furthermore, on the side where the push is dominant, the recessive trait will be a slight pull in the opposite direction. On the other side, the same motions of friction continue but from a completely reversed angle, what was mostly a push before the divide is now a strong pull, with a slight, recessive push in the opposite direction. Just using common sense, if there was not something resisting and causing some varied resistance in conjunction, there would be no such thing as a push or a pull. Therefore, there is pull in push and there is push in pull. Even though each side embraces its own opposite force, motion, and direction, the dominant trait wins.

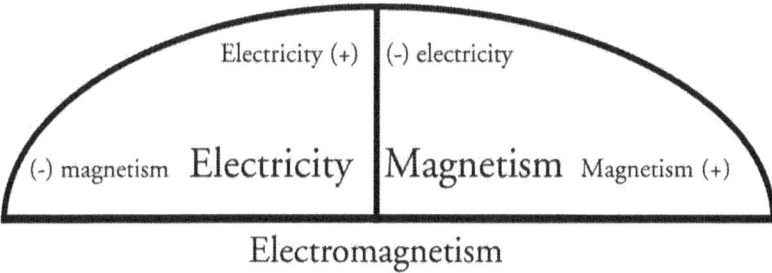

Electromagnetism

Electromagnetism is divided into two equal parts—electricity and magnetism—that are basically the same thing except in reverse. This unified force is what makes a simple motor run as well as transports the complex light and energy spectrum from the sun that sustains life on earth.

Electricity produces a magnetic field and a magnetic field produces electricity. On the side of the equation that has electricity as its dominant trait, magnetism will be its recessive trait, and on the other side magnetism will be the dominant trait, with electricity being recessive. As a result, there is magnetism in electricity and there is electricity in magnetism.

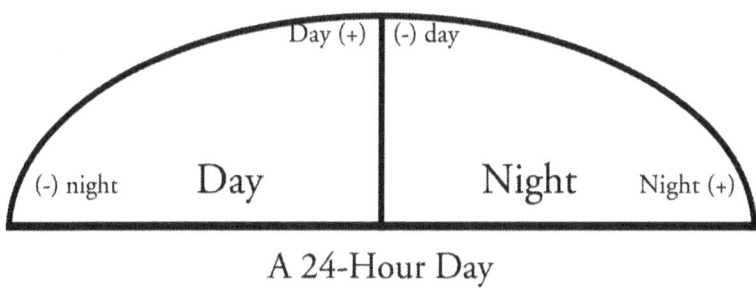

A 24-Hour Day

A 24-hour day is divided into two even 12-hour sections: day and night. They are basically the same thing except in reverse. The daytime is predominantly the result of bright light coming to us from the sun, with varying recessive shadows ranging from low light intensity to areas totally void of light. Without the aid of these recessive traits, the ability to see in the daytime would be anywhere from greatly impaired to non-existent. There must be contrasts of light in order for the eyes and brains to form visible images.

Nighttime is just the opposite. Its dominant feature is darkness, with low-casting, recessive beams of light coming from the moon and stars. This side's recessive traits help light up the sky at night. Therefore, illustrating the same consistent pattern as the other templates, there is night in day and there is day in night. Moreover, the day was ideally designed for waking and the night for sleeping, two other necessary, joint opposites.

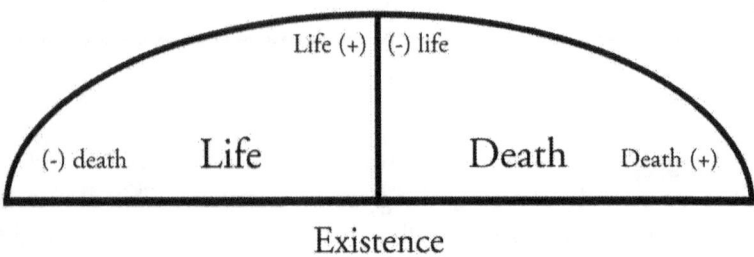

Existence

Perhaps the most important category to all of us would be the existence of life in its two interacting forms: life and death. When you ponder its mysteries, it can be rather confusing or frightening. However, when analyzed and set up in the same framework as all the other basic categories, it too can be discovered as having the same "genetic" makeup and predictable patterns.

The recessive side of life ((-) death) can be represented by the cells in living things that die within the dominant living process on a daily basis, a vital function we need in order to live. In contrast, the ever-living seed that is recessively hidden ((-) life) in a dominantly dying or dead organism is ever present, making these two opposites intrinsically linked and interdependent. Systematically, if there was no such thing as death, there could be no such thing as life and vice versa. Consequently and consistently, there is death in life and there is life in death; therefore, we are literally living and dying at the same time, because both sides are basically the same thing except in reverse.

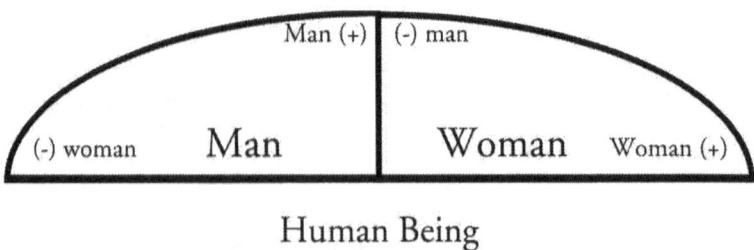

Human Being

The human being basically comes in two forms: male and female. Each is a duplicate of the other except in reverse: what is dominant in one is recessive in the other and vice versa. The dominant traits are designed for the purpose to give and the recessive traits to receive. Furthermore, each gender receives its own identity from the other. If there were no such thing as woman, there would be no such thing as man and vice versa. The more you learn about masculinity, the more you will understand about femininity and vice versa. Each fulfills, balances, supports, and compliments the other.

Both forms fit the full criteria for a human being, complete within itself. Each gender has everything the other has—masculinity and femininity, male and female hormones, etc.—only in proportionally reversed levels. Therefore, each gender's reproductive organs are the reversal of the other, making the two opposites a perfect magnetic match (discussed in an earlier chapter). Falling in line with all the other templates, in this comparatively important classification, it is also evident that there is woman in man and there is man in woman.

Thought Patterns

The above graphs cover a great deal of the basic fundamentals in our world, but what about the world within us with which we consciously commune daily? What about our thoughts, the medium we use everyday to interpret our surroundings, to determine our relationships, achievements, or failures, and which cause our pains and pleasures? Because they are intangible, are they beyond measuring, interpreting, and controlling? Or could they possibly fit the same mold and criteria that constitute real organization and real understanding? Without abandoning the same structure, mathematics, and rationale that have taken us to this point in successfully observing the reality and format of our world, these next templates put the mystical realm of thinking on the chopping block as well, for dissection and illumination.

If we have discovered that all life's basic, fundamental categories are structured within the orderly arrangement of a paradoxical continuum, then we must reason that thinking, and how we learn would be no different. Anyone could acquire knowledge in many different fashions and have sophisticated, mind-boggling speculations, and we do; but real understanding must fit the proper format that cross-examines and verifies itself in a way that only a pair of opposites can do. However, does it work the same for this imperceptible and elusive category? Can the process of thinking be laid out and observed in the same fundamental terms as all the other units?

It has been very hard for mankind to accept the ugly and dark side of the equation and understandably so. This is most likely the reason for the brick wall or the roadblock that has stopped many truth seekers in their tracks because they could not go any further down that dual road, for it inevitably leads to the moral issue of good and bad, or good and evil, thereby causing a complete rejection of the whole plausibility of opposites, not being able to accept both sides of the equation. This is the adductive, theoretical result that comes from viewing the world in dual terms and paradoxes. In the philosophical or logical procedure of adduction, one idea adds to and supports the other. Therefore, if one cannot personally accept the reality that the awful side is inherently designed and automatically reactionary, then the most natural and common thing is to deduce the duality concept rationally, calling it a contradiction instead of a paradox.

However, given enough time, experience, and evidence, it can be proven how and why this unacceptable side exists alongside the one we do like. We can choose to go along with the common consensus or common sense of rejecting such a seemingly ridiculous idea, or we can analyze it with real scientific tools and facts, thereby deciding for ourselves. The following controversial graphs concerning our thoughts continue in the same outline form as the previous templates, using the exact same guidelines and measuring tools of the paradoxical continuum.

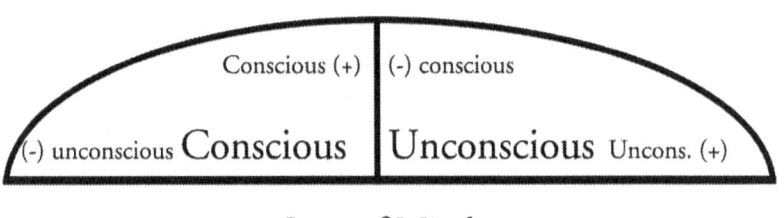

State of Mind

The state of mind is divided into two basic halves: conscious and unconscious. Again, they are the same thing except in reverse. The conscious mind is the dominant side when we are awake, with its recessive trait coming to its aid in a very subtle and hidden way. This was thoroughly explained in Chapter 1. In the opposite half, with the unconscious mind being dominant, as when we are asleep, its recessive trait is what allows a small amount of consciousness for a person to wake up to motion, touch, or respond to sounds, etc. and, perhaps, to carry on an unconscious conversation with someone who's awake.

Consequently, we find the same transcending pattern applicable to this category as well, in which there is unconscious in conscious and there is conscious in unconscious, harmoniously working together to create our state of mind. Neither half is an isolated characteristic, as some might think. Each serves the other in a recessive and cooperative state.

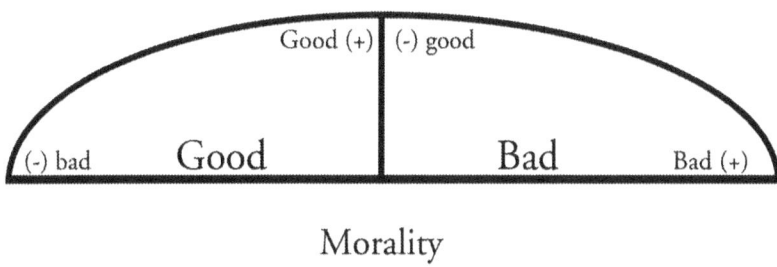

Morality

"There is so much good in the worst of us and so much bad in the best of us that it hardly becomes any of us to talk about the rest of us" (Anonymous). Good and bad are directly linked together also because they are the same thing except in reverse. Bad can't be just anything; it has to be the exact reversal of what is considered good, just as you would have to be educated in the laws of government in order to be declared or to know you are an anarchist. One must also understand capitalism in order to be labeled a communist or to know the fundamentals of Christianity in order to be a Satanist. Understanding cannot be isolated. No matter what it is, it is linked to and clarified by its opposite half: to know one, you must know the other.

Another example is counterfeit money. In order for it to be a perfect counterfeit, it cannot resemble just anything; it has to appear as real money in complete detail. But, just like looking into a mirror, the reflected image is in reverse. A good counterfeit is a perfect replica, but it is the exact opposite of what is real. Therefore, counterfeit money will be as similar as it is different, because it is completely reversed in all its qualities and applications. One is real, the other false. One has a legitimate source, the other illegitimate. One is intended to give, the other to take. One is earnest tender, the other deceitful. Moreover, one verifies and cross-checks the other.

There are always pros and cons of any situation as well as positives and negatives, goods and bads. However, they have an order, and that order is no different than the same dominant and recessive patterns found in all the other basic templates of life mentioned here. Therefore, this category is no exception, and we must get past our preconceived ideas in order to look reality in the eye and finally see it in the same light as all the other units. For, there is bad in good and there is good in bad, just as there are adversities in blessings and blessings in adversities. Seeing mathematical duality in the moral issues is a higher level of perception that a lot of people are incapable, unwilling, or fearful to explore, but it is necessary in order to get past the present erroneous dogma and enlighten us toward an advanced and balanced viewpoint.

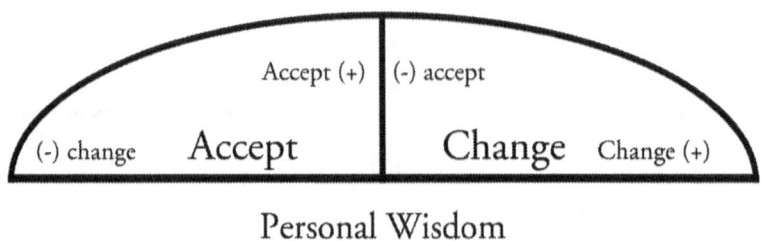

Personal Wisdom

"God grant me the serenity to accept the things I cannot change, the courage to change those things I can, and the wisdom to know the difference." The Serenity Prayer, often attributed to Reinhold Neibuhr, has been a famous and favorite one over the years, not by accident but by design. These words not only describe the magic formula to success but also, as anyone can see, are subtly arranged and mathematically structured in a continuum. Acceptance and change are really the same mental actions, just in reverse form. On one side of the equation, *accept* is the dominant trait and *change* the recessive, whereas *change* is the dominant and *accept* the recessive on the other side. Therefore, there is change in acceptance and there is acceptance in change.

So many of us could spare ourselves a lot of anguish and frustrations if we would only learn to apply this simple but very effective philosophy. Instead, most of the time, most of us keep trying to change the things we cannot change and choose not to change the things we can or should. Knowing and applying the difference is indeed wisdom. Believe it or not, it reminds me of playing poker, even though I never learned how. It may be unwise to play poker, but there is wisdom in poker, because the following lyrics describe well a key ingredient in the decision-making process: "Know when to hold 'em. Know when to fold 'em." A valuable lesson and balancing act for all of us is to know what we should accept and what we should change. And there's no better way to do that than to have all our choices lined up in a cross-examining springboard of balanced opposites, pros and cons, positives and negatives, and, ultimately, what works and what doesn't. Knowing one will help us considerably in evaluating the other.

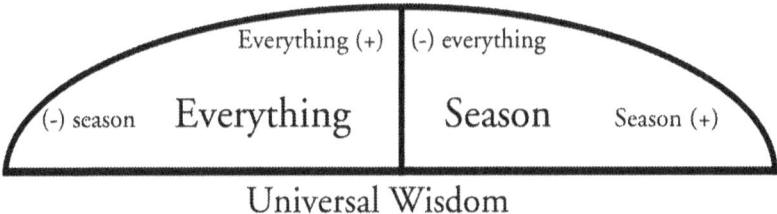

Universal Wisdom

The following quotation from the Bible remarkably reveals the same universal organization that is found in the above templates and that exists in all of life's personal and human experiences as well:

> To everything there is a season and a time to every purpose under the heaven. A time to be born, and a time to die; a time to plant, and a time to pluck up that which is planted; A time to kill, and a time to heal; a time to break down, and a time to build up; A time to weep, and a time to laugh; a time to mourn, and a time to dance; A time to cast away stones, and a time to gather stones together; a time to embrace, and a time to refrain from embracing; A time to get, and a time to lose; a time to keep, and a time to cast away; A time to rend, and a time to sew; a time to keep silence, and a time to speak; A time to love, and a time to hate; a time of war, and a time of peace. (Ecclesiastes 3:1–8)

The first sentence of that paragraph is none other than a recognizable equation. The first half is just another way of stating the second half only in reverse order, for everything is synonymous with every purpose as well as all (the contained/the parts), whereas season is the equivalent of time and one (the container/the whole). What this declares, in simple language, is that everything here in this world, in the manner we know it, has a purpose and a time slot that is arranged in a seasonal or timely manner. For example, our seasonal weather is an organized, timed structure divided into two basic, extremely dominant opposites, summer and winter, with spring and fall being their recessive and milder attributes. Hence, it could be said there is winter in summer and there is summer in winter.

However, the main connotation in this parable is not weather, but something far more fundamental. It is basically stating that everything (no exceptions) is contained in a timed sequence of events such as a season. The continuum of time happens to us all and doesn't stop; it continuously marches on alongside things and events. And in the opposite pole, everything and its purpose in that arranged time period (season) is just as important as any other occupant, each being relatively connected and perfectly unique at the same time. This phrase could also be described by another template of life: all in one and one in all. But perhaps even more important is the follow-up of that scripture, explaining in detail the times of our lives (e.g., "a time to be born and a time to die"), the emotions, activities, and experiences we all share in as human beings, and how each one is legitimate and individually as well as collectively important. Thus, as this scripture portrays, everything (every purpose/all) is in a season (time/one), and a season (time/one) is in everything (every purpose/all).

Last, but not least, there is one more revealing truth in this declaration. Remember that in the all in one and one in all paradigm, the recessive one in the second half of the equation is essentially a skeletal blueprint of the dominant one in the first half. Since all the examples given in this scripture to describe the different kinds of times we experience are set up in a two-sided continuum of opposites, then this means the overall (whole) arrangement, one (season), is in the form of a paradoxical continuum as well. Consequently, in all rationality, it can be said that everything is in a continuum, and a continuum is in everything. As a result, everything has order, purpose, time, and place, and that order is dictated by and built upon paradoxical opposites that are divisible and multiplied throughout.

The next chapter takes the above research from organization to application. The graphs up to now have displayed the universal arrangement and basic structure of all units. The following continuums will instead focus on the transactions and inner workings of the paradoxical formula. They will be viewed more from a fluid and progressive standpoint: how things progress, mature, and are learned, instead of just the demonstrated order by which they are already structured and properly function. Then perhaps we can decide from our end of the spectrum if we can apply the same methodical steps and organizational layout to the learning process, getting the same thorough assessment and conducive environment that the paradoxical continuum provided in all the other templates of life demonstrated so far.

6

The True Learning Curve

Once we understand more about how our thoughts and emotions work, their basic structure plus the functional laws that they obey, the better we can have control over them. It is really no different than understanding how a motor runs. If we understood the mechanics as to how it's put together, what makes it work, and why, we would be in a better position to know how to operate, fix, and maintain it. The following graphs and exercises will prove that our thinking and learning processes work according to the same paradigm found in all of life's magnificent entities. For, the all in one format requires that all things function, basically and generally, under one set of rules without exception.

In a previous chapter, experiments were described in detail about how our sensory awarenesses and assessments are acquired. The following categories that deal with our thoughts and emotions will prove that they exude the same laws that underlie all the others. For example, in the case of experiencing the various awareness levels of black and white, outlined in a former chapter, we could see that the learning process was subject to the exact ratio and proportions of the two opposite components. When we transfer the same mechanics to the personal side of our lives, we will see, surprisingly, that it all works in the same oppositional manner.

The intermediate, interdependent variations encased between the two opposite bookends of a unit are typical of learning stages, whether the category is tangible or intangible. This basic standard crosses all barriers. It isn't confined to just physics, etc. For instance, the more you learn of one thing, the more you learn of its prerequisite opposite in equal proportions, which is relevant to one of Isaac Newton's laws: "For every force, there is an equal and opposite force." Without a doubt, the harder you hit a ball with a bat, the further it will go in the opposite direction from which it was thrown. The amount of energy an atom can radiate is the same as the amount of energy it absorbed. The stronger the electrical charges in binding molecules together, the harder the material becomes. The mathematical formula is the same regardless of the category.

The interplay of opposites, regulated by this law, causes the same motivation in emotions as it does in physical energies. In all reasonableness, nothing could be more right than to have enormous love for what is good, matched by enormous hate for what is wrong. If a person doesn't much care if things are wrong, naturally, he or she doesn't much care if things are right. One motivation activates an equal and opposite response. Justice also requires both intensities. Whether we like to admit it or not, tremendous love automatically consists of tremendous hate; you cannot have one proportion without its equal and opposite attachment. Same as the batted ball, the more you love something, the more you are going to hate what opposes it. The more you love a certain way of life, the more you are going to hate the reverse (elaborated in the next chapter).

This is the stage of learning in which most people stop the examining process because it doesn't personally make sense to them, and they begin impatiently to interject their own shortsighted and biased opinions into the process before it has completed itself. However, if they would just let the final stage unfold, never straying from the complete path of its learning field, the truth will become evident and obvious. Many philosophize that the absence of love would be its ideal opposite instead of hate. However, nothing supports that theory mentally, physically, or mathematically. It has been proven that an opposite is the same thing except in reverse. In the case of absence, nothing wouldn't cut it, for there has to be something there for it to be a reversal or an opposing component.

The world in general wants to eliminate the unfavorable side of the equation: bad, ugly, pain, anger, guilt, and hate, but people find that they can't. Even if we invalidate it in our minds and try to get rid of it, it just resurfaces again, rearing its ugly head somewhere else, such as crime that emerges from freedom or the adversities that come from blessings. Instead of burying our heads in the sand or trying to recreate nature the way we think we would like it, the wise and rational thing to do would be to admit we can't get rid of it and frankly accept its right to exist. Then we could study it, uncover its laws, understand it, deal with it, and learn from it.

In all honesty and reasonableness, one has to concede that the following events and emotions are really no different than Newton's physical law of balanced forces, and each one obviously obeys the same laws of opposites:

- The greater the suspense, the greater is the mystery.
- The greater the villain, the greater is the hero.
- The greater the search, the greater is the discovery.

- The greater the anticipation, the greater is the disappointment.
- The greater the denial, the greater is the desire.
- The greater the sorrow, the greater is the happiness.
- The greater the fear, the greater is the courage.
- The greater the war, the greater is the peace.
- The greater the obstacle, the greater is the triumph.
- The greater the desperation, the greater is the hope.
- The greater the testing, the greater is the faith.
- The greater the wrong, the greater is the forgiveness.
- The greater the love, the greater is the hate.
- The greater the pain, the greater is the pleasure.

The feelings of pain and pleasure are perfect examples of this equal ratio. Explore the hypothetical possibility that, for the first time, you experience a little pain and then, and only then, do you realize you weren't quite so bad off before the pain arrived. Just like Adam and Eve, you did not realize how good things were until things turned bad. Instead of ignorance being bliss, we can instead be oblivious to the fact that we are happy if the other side of the equation has not been introduced or activated. When we experience terrific pain, then we realize more fully how wonderful pleasure is. In conclusion, you do not learn the full extent of something until you have experienced the fullness of its opposite. Consequently, the learning processes are equally balanced between its two contrasting poles just as they are in physical reactions.

The graphs below will demonstrate how the two opposite sides of a continuum work towards creating a learning curve for us—true lessons in how to reach successful harmonious balances in our lives—as well as teaching us how to develop an authentic, automatic monitoring system to maintain that progress and success. Again, this all-in-one package is the only thing that renders a victorious outcome, no matter where you use it. The math can be applied anywhere. It is the guideline, measuring tool, and the perfect learning curve we've all been looking for but can't find because it has been buried in a mountain of errors, prejudices, fears, and shortsightedness.

Creating Successful Equations

The graphs in the previous chapter exemplify consolidation and representation, the full embodiment and equalization of the two halves; whereas the following life templates represent progression and successive steps towards developing the end result—the successful completion of a unit, its course or cycle—and, ultimately, maintaining it. Finally, knowing the definition of success (besides love, the most undefined word in our vocabulary), how to achieve it, where to pinpoint its exact, ultimate location, plus the outline of exercises needed to adjust constantly the balancing throttle that sensitively fine-tunes what was achieved, and keep it running smooth, is indeed success.

Instead of displaying the perfect scenario, how each opposite is the recessive component on the other's dominant side, being the same thing except in reverse, the following illustrations depict the two extreme, dominant bookend components as the two undesirable ends of the overall spectrum. They represent exclusion, extremism, and incompletion. However, as they move towards the middle of the continuum with time and maturation, they balance their own magnetic poles in each one's individual spectrum by enveloping their recessive trait, thereby filling up their half of the equation. When this is done and they meet each other halfway at the middle line of division, something magical happens. This center, pivotal point where the two opposing equal sides come together in fullness is what creates the peak of perfection, the summit, the epiphany, true success, a happy medium, harmony, serenity, a peaceful calm, the best of both worlds, or a balanced viewpoint, as the following illustrations display.

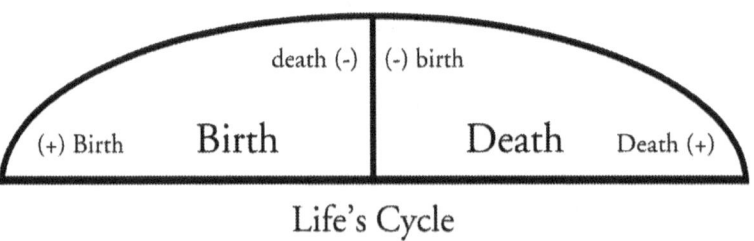

Life's Cycle

There is a beginning, a short peak, and an end to all living things, just as there is a past, a present, and a future in the measurement of time. All things grow to their full potential in the first half of the equation, culminating and maximizing to their peak performance or state of being. This is the summit, the perfect state, and the middle of the continuum, which is a very short span of time, like the blossoming of a flower; from that point on, it quickly enters the second half of the equation, which is its opposite, the declining or decaying process. When the first half crosses the fine line onto the other side, it does encounter life firsthand, but in a recessive and negative form, as it begins to decay. Then, as it travels to the dominant end of that spectrum, it dies and completes the cycle of life and death.

This cycle requires both opposites to exist for there to be a peak or a point in time of maximum perfection. If there were not two opposite halves doing the exact opposite thing, one growing and one dying, this evolutionary process could not take place nor would there be a status for perfection. Nor would there be such a thing as a beginning and an end, birth and death, spring and fall, and, most unfortunately, no mentally differentiating discernment. Furthermore, neither extreme, the beginning of life or the end of it, is the ideal state of being, but it is what they produce when they come together at the middle, when the growing cycle ends and the decaying process is about to begin, that creates the pinnacle peak of life. For instance, the descriptive phrase "over the hill" describes quite well in continuum form the growth and development of human beings and the ideal age that is a relatively short period of time before it switches to the other side, beginning the decline.

As for the final characteristic housed in this unique paradox, the learning feature is one of its most helpful. By having foreknowledge of and experiences in life's continually repetitive cycles, a beginning, a middle, and an ending, creating a curve of information—its incline, peak, and decline stages—we have a clear standard of measurement and a learning process for timing, like the past, the present and the future. We can look at any segment along this path and pinpoint its current stage of growth or period of time. Without this relativity and standard, there would be no such thing as timing in the way we know it. As far as intensity and expertise go, the learning stages are also linked: the more we learn of each stage, the more we learn of the others proportionally, each identifying the other.

In conclusion, because of these repeated patterns of opposites with beginning and ending cycles intrinsically existing in our lives, we can have confidence in what we learn from them by letting the consistencies of nature be our guide, creating self-reliance because the environment is us. We cannot be separated from it, for we are a part of and a result of it. Without its reliable and recycling laws, we would be nothing, for we could not exist otherwise or think soundly. Therefore, the more in sync we are with nature, the more we think in patterns of cycles the way our environment is designed, and the more we will think in terms of continuums. The more we think in spectrums and continuums, the more organized, complete, and authentic will be our thoughts.

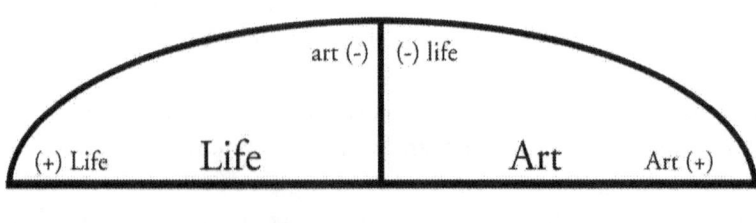

Created Scenario

Many subjects can create many different categories, depending on their diverse purposes and applications. Creation is one of them. For example, one unique category it generates can be divided into two basically opposite creations—life and art, that which is apparent or real and that which is projected or unreal, being the same thing except in reverse. This really shouldn't be too surprising, for these crossovers are often paired off in a continuum that has created a popular question: Is art imitating life or is life imitating art? As usual, most people mull over this dilemma and think it should be one way or the other, but of course the true answer is that both do. These two opposites have influenced and shaped each other in an interdependent and orderly fashion that universally flows in all transactions. Each is a genuine spectrum of its own, but each of these creative influences requires the other to do the opposite of what it is doing in order to complete this paradox, the way we live our lives. It may appear strange, but it is the phenomenon that has shaped our world.

There is art in life and there is life in art when each completes its own spectrum of opposites. When people see something in nature at its best they will say: "It looks so perfect, it looks artificial." Then when they see beautiful art, such as an artificial flower, they say: "It's so perfect, it looks real." Each reaches perfection by surfacing its recessive opposite. When art and life are at their best, they can be mistaken for the same thing, and, ironically, it is that deception (the transcendence) that moves people emotionally. The best novels are those based on real stories with a fictional takeoff. And, in reverse, a far-out fictional story must have some semblance of reality to make it halfway believable. For a book to be a best seller, it must incorporate within its broad base its opposite in a recessive role. If people wanted to only have history influencing their lives, they would just read history books.

As far as life imitating art, one of the best examples is science fiction. It is hard to say how much effect science fiction writers or predictors have had on our real lives, fashioning our societies somewhat according to these preconceived ideas and forecasted inventions. Needless to say, our world would have been different without that imaginary influence. For instance, if there were such a thing as Martians, there's a good chance we wouldn't recognize them if possibly encountered because we already know what they look like. Thus, we have shaped our real world and minds, in some measure, according to our own artificial designs, just as surely as life's real designs are essentially imbedded or represented in our world of art and make-believe. Imagine what the Christmas scene that people live out every year would be like if Charles Dickens had not been born or literature from the 1800s did not exist? The glamorous, glory days of Hollywood in the 1900s created untold movies reproducing real life and history as we know it to be, along with a fictional romantic twist, thereby developing a situation in which art imitates life. But who can figure how much of that subtle, highly underestimated fantasy part has fashioned real lives beyond the silver screen, whereby life imitates art, that which was unreal.

One of the biggest and most unanswered questions, besides what is love and what is success, is what is art. People are beginning to think that anything can become art, even if it's mud splattered from a car onto a canvas. But that is not true; it cannot be just anything. A true definition of anything has to be the reversal of the other half of its equation. Therefore, art is the exact opposite of life and vice versa. And in order to know what art is, you would have to know what life is. They are intrinsically and consciously linked together. Art comes from the word artificial and rightfully so. Life is real and obvious, whereas art is unreal and deceptive. But the guideline by which each is tested or verified is not just any standard. Just like the counterfeit money, even though it is a fake, it has to be easily mistaken for the real thing. All the genuine fine arts that are successful are involved in imitating life, but their creations are not real. When something is done so well in the field of art that it transcends the mind into thinking it is real even though it is not, that is the fantastic feature that makes it art. In fact, the greater the transcending and deceptive effect, the greater is the art.

Take for example a great painting. It has to do one of two things: (1) be so good that it can be mistaken for a photograph, or (2) capture a mood or surrealistic feeling so effectively that it can transcend us in time, in place, or in emotions, the way music does. It is not so much the skill and accuracy that make both kinds of art great as much as it is the unreal or artificial virtual reality journey it takes us on, even though those skills are highly required, because it is only when it is done skillfully that you can become oblivious to the reality and can get caught up in the fantasy. If the bubble is burst by a blunder, so is our trance.

If a ballerina did not pretend that her gravity-defying leaps were effortless, with finesse and graceful poise, or that standing on the tip of her toes in a very cramped ballet shoe did not hurt, it would not be art, and we would not enjoy the fairy tale-mood producing scenario. If a dancer had the exact same skills or maybe even greater, but he or she frowned, grunted, sweated, or, forbid, winced in pain, it would be an athletic feat but not an artful event. The same transcending achievement comes from reading a good story. Its success lies in completely transforming and sweeping the captivated reader into a visionary, colorful setting, event, and circumstances with such intensity and vivid imagery that he or she is oblivious to the fact of being in real time, just reading black opaque etchings on white paper that forms symbols, words. Ah! the magic of another pair of opposites, good literature and a seasoned reader.

To sum it up, there are two wrongs, with neither extreme bookend being the ideal situation: living in a real world without dreams and living in a dream world without reality. When artful expectations and dreamy imaginations do not conform to reality or when reality cannot live out its dreams, it is not a good match. However, when these two opposites, synthetic fabrications or expectations and reality, come together at the middle, meeting each other halfway and equally imitating each other, it creates a sensitized and effectual union. Our artificial craftiness, idealistic imaginations, and speculative projections do guide and affect the way we live our lives for the good and for the bad, whether true or false, creating the kind of environment we have, but they can only be successfully materialized if they are grounded and attached to the laws of nature, life that is real.

Personal Equations

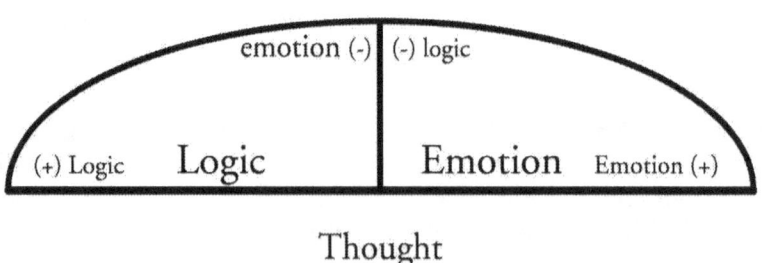

Our thinking is regulated by only two basic kinds of thought: logic and emotion. They are, needless to say, opposites, being the same thing except in reverse. Both are thoughts, but logic is basically reasoning without feelings, and emotions are basically feelings without reasoning. Neither extreme nor half of the equation is the ideal attribute. Obviously, we need both traits in order to interact properly within ourselves and with our environment. Logic without emotion is just as bad as emotion without logic. One bookend would produce a cold and calculating person, and the other a mindless idiot with emotions running amuck. That we need both is apparent to most people, but there is always the proverbial question: "But where do you draw the line?" The answer should be simple and to the point, but many consider it mystical and unanswerable. Naturally, it is the line drawn in the middle of a paradoxical continuum. These two opposite thoughts, when equal, create a balanced and working relationship.

However, knowing and doing are two different things. Knowing the correct formula is vital, but applying that knowledge is the catalyst that can only bring success. Therefore, the question is this: How do we actively balance our thoughts between these two oppositions? Since our thoughts usually determine and control our actions, we must begin by systemizing and organizing our thoughts. There is a fractional reason why knowing and doing do not always add up. To be able to consummate emotion and logic equally, we must consciously put in the right ingredients with the right measure, just like a recipe, in order to get the right results, remembering that too much of anything is too much and too little of anything is too little, no matter what it is. If our minds dictate and control our actions, then someone or something should be organizing and controlling our thoughts. Of course, that someone could and should be ourselves, but how can this be accomplished when the intangible act of thinking can be such an elusive and mysterious thing?

By using the general transfer of basic knowns to unknowns mentioned before, I have found that the best way to illustrate and understand the effects of these two continuum bookends working together is with a conventional clock. Straight up the middle, 12:00, is the dividing line separating the two opposite halves. The morning side, 6:00 to 12:00, represents logic, and 12:00 to 6:00 in the evening represents emotion. If the hour hand moves over to 3:00 in the afternoon, then the logical side has passed its ideal halfway mark and is invading the emotional side, making this person too calculating and rather cold, thereby impeding and inhibiting his or her sensitivities and feelings. And, in reverse, if the hour hand stops at 9:00 in the morning, then logic is being squashed out by the expanded emotional side, with very little room to think logically, making this person listen more to his or her heart than to his or her head, which gets most people in trouble. This simple, picturesque measuring tool helps tremendously as a quick mental reference when balancing other categories as well.

In conclusion, the right ingredients for a sound, temperate mind is half and half. In addition, the right monitoring system is a constant mental thermometer, always focusing and checking for the correct balance, remembering that too much logic is just as bad as too much emotion and that too little logic is just as bad as too little emotion. Making these two characteristics equal is the magical key not only for knowing that something is the right thing to do or to be, but for having the ability to carry it out as well. There must be a conscious effort to make sure they are balanced at all times. This will provide the capability of making sensible, loving choices and logical decisions that are humanized. Logic that is enhanced by caring, generous, and sensitive feelings is just as satisfying as emotions that are guarded, guided, and checked by logic; each cross-checks and controls the other. Here again, dominant and recessive roles play their important paradoxical game, perfecting the unit.

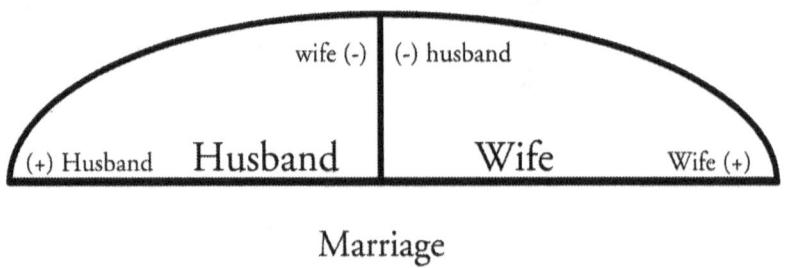

Marriage

Perhaps, the most important union that has existed, does exist, and will exist is that of man and woman forming the marriage bond and becoming husband and wife. However, this enduring merger has unfortunately suffered extreme relationships, creating a myriad array of dysfunctional arrangements that have spanned a wide spectrum, anywhere from servant and master, to sex wars, to non-binding existence. How these two opposites should coincide has been a subject of inquiry and mysticism since history began. Even today, this unsettled issue fills the bookshelves in a variety of prescribed solutions and protocol, developing an ever-expanding supply and demand situation: those seeking answers and those giving advice, which is a never-ending process. As a result, it is confusion running rampant, in which anything goes concerning the proper and productive roles each should have, causing even more uncertainty, disorder, and frustrations in the home.

The answer should be a simple one. But, then again, it seems the simplest and most basic questions are the hardest ones to answer. Besides, the threat of being politically incorrect can make one wary of commitment. It is hard to go against the grain when obscurity reigns and absolutism is taboo. On the other hand, if we choose to be logical and scientific, in spite of what is popular, we must ignore the present trend and follow the path of consistency, patterns, and rules of nature instead. To disregard totally the sound, consistent, and universal laws demonstrated in all these categorical, paradoxical units studied thus far would be about as unscientific and dishonest as one could get. Therefore, trying on for size the paradox template and seeing that it fits this category, as it did all the others, shouldn't be too big a surprise at this point.

In recounting the study of man and woman, it was discovered that each is a complete human being with everything contained in one existing in the other also, except in reverse proportions. Having already catalogued their even match up with anatomy and sex features in previous chapters, this classification deals primarily with the roles of husband and wife. Being that these two halves of a marriage are opposites, they are automatically the equal bookends of a continuum designed for this institution. Therefore, the leading question is this: If they are equal opposites, are their roles in the marriage unit equal as well? Should it be, as some suggest, an arrangement where one leads and the other follows? Should one be the boss with the other subservient? Or should one be the commander-in-chief and the other one next in charge, like a pilot and copilot? Since the general public has yet to find a standard by which to measure and guide its evaluating processes, then there are probably as many different answers and suggestions out there as there are people or groups. However, via the rejected cornerstone, the understanding of this subject, like all the others, simply and categorically falls in place.

In spite of the popular vote, status quo, or sophisticated speculations, using the modest elementary clock analogy works remarkably well for this continuum, as it did for logic and emotion. Hardly anyone questions that each plays a role, but the questions are which one's major and which one's minor? and where do you draw the line? There again, it is a simple monitoring process when you envision each being half of a clock, a unit whereby you can measure two equal opposites balanced side by side, with each opposition containing both major (dominant) and minor (recessive) roles. If the husband is hypothetically on the left side of the clock, 6:00 AM to 12:00 noon, and the wife's position is on the right, from 12:00 noon to 6:00 PM, it is obvious these are equal slots. If the husband takes his role all the way over to 3:00 PM, dominating and inhibiting the full support and productivity from his wife, he denies her side from becoming a fulfilling role. As a result, this marriage has shortchanged its full potential, creating a warped and unbalanced arrangement, a predominantly male institution. Of course, the same result comes in reverse from a wife operating and controlling beyond her rightful territory, stripping him of half his power and rights, creating a predominantly female institution.

There is another dysfunctional match up that is often too common in marriages, and that is when one side fails in its responsibilities to fulfill its role with an unenthusiastic or detrimental attitude, leaving a gap that the other must fill. The relationship of husband and wife is quite similar to a very common physical law: nature hates a vacuum. If the left side of the clock stops at 9:00 AM, leaving a gap to 12:00 noon, then that void is usually taken up by the right side, overextending her energies and duties. And there are just as many cases where this situation is reversed: a gap left between noon and 3:00 PM, causing the husband to exhaust his resources because she neglects her input and responsibilities. Of course, it is obvious that, if one side chooses not to take up the slack or is incapable of it, then that breach and deficit usually breaks the bond, causing its dissolve.

Consequently, the ideal arrangement is one that works like clockwork. Instead of rivals for power and importance, each half of the marriage continuum works in a cooperative give and take, ebb and flow, interactive partnership, each contributing his or her full input, potential qualities, and sovereignty to the unit. Either extreme—feminine control or masculine control, foreign or domestic—produces an incomplete fraction, an unbalanced equation, and less than pleasing results. Of course, there is always the inept and narrow-minded statement that always pops up when discussing this subject, such as the following: "Well, someone has to be the boss." That is true, but, when roles and duties are divided into two major opposite and equal categories, there needs to be two opposite bosses.

To this day, there are many people, who still cannot get past the persistently stubborn philosophies from the days when monarchies and dictatorships reigned supreme, now rendered antiquated by a world where equality and democracy are king. That is why the simple, logical, scientific process called mathematics plays such a significant role in solving such controversies. Just do the math: $1 + 1 = 2$, an equal partnership, or $\frac{1}{2} + \frac{1}{2} = 1$, a whole. Both are bosses and followers of each other in the roles they are equipped for, in positions that are dominant but contain a measure of each other's capabilities and traits, each leading and following the other.

To put it in perspective, yes, jobs and buildings, goods and services, business institutions and governments are extremely important in maintaining our existence and society. However, are all these entities and necessities more important than the number one commodity a nation produces, which is its children and its posterity? Do we have children so that we can have goods, services, offices, buildings, schools, and governments, or do we have goods, services, offices, buildings, schools, and government so that we can have children? Can anyone honestly say that a business executor's or a government official's position should be more important than the humble upbringing and nurturing in the home, of a country's most important product that carries a nation onto the next generation? The answer should be obvious and mathematical: a whole or continuum consists of two equally important halves whose roles are the exact opposite of each other yet are the same thing except in reverse. When we shape life according to its proper and paradoxical designs, we have harmony, balance, perfection, purpose, and clarity.

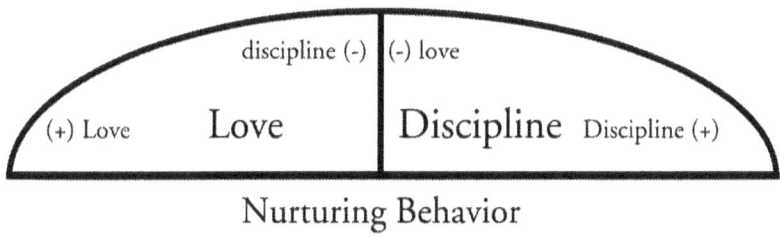

Nurturing Behavior

Another great supply and demand situation that is stockpiling bookstores is how to raise children or how to deal with their problems, creating a wide range of varied solutions with dysfunctional and ridiculously unfounded philosophies, thus causing a bad situation to become worse by an ever-increasing explosion of soaring speculative remedies that never touch the ground.

However, could the proper guideposts for this subject, like all the others, be just another paradoxical continuum with simple, specific rules of conduct towards rearing and nurturing children? Could this little prototype have possibly been our panacea, providing the world with a road map to good parenting had we not avoided facing the equation that brings us both good and bad, love and hate, blessings and adversities, punishments and rewards? Could one of the most importantly endearing categories and commodities in people's lives perhaps be managed in basic mathematical transactions? And, last, could we as a people or nation have done better than to go from the extremism of strict, cruel disciplinarians to excessive, loving liberators in just a few short generations? Could justice have been better served had we not jumped from a narrow-minded tunnel vision, from a one and only answer for behavior control to no answer?

The truth has to be an astounding yes, when there is absolutely no situation that does not benefit from a balanced position, remembering that the conditions for balance cannot exist without two equally opposite oppositions. This equation is by far no exception, for there are ultimately two wrongs in this classification: (1) to show only love and kindness towards children, no matter what the situation calls for, with no boundaries, no consequences, or punishments, thereby affording them no responsibilities for their actions, and (2) to only discipline or punish children, like a military drill sergeant, offering no love, kindness, or conditional rewards, just more punishment. Needless to say, the perfect balance between these two extremes would be the ideal condition. Not even an uneven or incomplete fraction would be satisfactory, because an imbalance creates warped arrangements and, even worse, flip-flopping from one exclusive pole to the other, brings about serious inconsistencies and dangerous results, sending extremely mixed messages to those being nurtured.

Eventually, I developed a phrase that has served me well when working any equation, especially one concerning children, and that is "working at both ends at the same time." Like any pair of opposites, each side should work together in unison like a seesaw, consistently and instantly at the same time. If one side of a seesaw does not come down, the other cannot go up and vice versa, and the faster, more extended the interactions, the better is the ride. Another good example that demonstrates this development is a memory I have of trying to contain a couple of newly acquired dogs inside an area fenced off with an electrically hot but safe wire less than a foot off the ground. At first it did not work. These young dogs kept repeatedly crossing through the wire, receiving a considerable shock each time. At first I was confused as to why they put themselves through this torture.

Eventually, I realized the pain must have been worth it or it equaled the anguish of being isolated and penned in. As a result, I decided to try working at both ends at the same time. We put a little doghouse in the middle of their yard, brought them tastier food, petted them and stayed for a while, smothering them with affection and comfort. And, low and behold, it worked; afterwards they never crossed the fence wire again. This time, logically and mathematically, they had an equal reason to stay, because the good evenly matched the bad, creating a whole and complete equation: the reward matched the punishment.

Thus, this balancing act between love and discipline not only works well on pets and animals but, of course, on children or anything else that requires proper and perfect nurturing. One good example of this achieved balance is regarding the subject of respect, which is such a crucial element to child rearing. There are two wrong bookends: (1) parents demanding respect from their children when they themselves behave disrespectfully, and (2) respectful parents who don't expect and demand respect from their disrespectful children. The ideal equation, of course, is respectful parents demanding respect from their children. If they are trying to do one without the other, it most likely won't work. First of all, the raw reality of life is that children will not act respectfully on their own if they are not forced to, no matter how respectfully parents act. Second, once it is demanded of the children, it is much easier for them to reciprocate that respect if they have a role model to go by. Not only can they comply better but also they will respect the way their parents live a respectful life. Doing what they see done is more highly influential than just doing what they are told. Therefore, both sides of the equation need to be activated and equally so.

Raising children today is probably the most difficult, grueling, and heart-rending responsibility there is. However, the task can be a lot less confusing, frustrating, and complex when we have such a valuable tool as the paradox, which discerns and monitors our methods and decisions so that we know the ideal environment can be just the simple, unwavering goal of keeping the pendulum resting in the middle of the continuum. There is no more guesswork as to how much love and discipline and how much freedom and restraint are needed or, for that matter, whether there should be any discipline or restraint. Then, finally, success will be easily recognized and child rearing ultimately rewarding when all the hard work and difficult times equally produce good times, fun, blessings, and the lasting enjoyment of watching your precious children grow up.

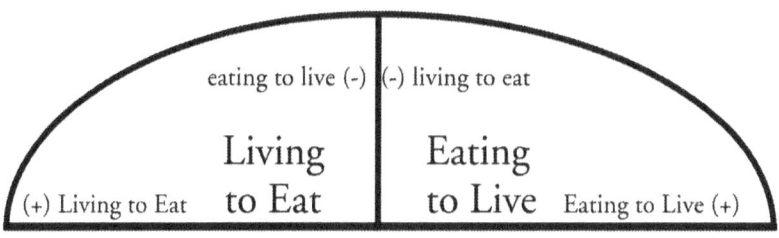

Consuming Food

The discerning continuum tools can give us an instant, broad-based picture of any situation, even when we are trying to manage some of the simplest things in life, such as eating, an ordinary act that you normally wouldn't put much thought into, much less think it to be a mathematical problem. But, just looking around us in this overly indulgent society, it is easy to see what happens when there are no bifocal viewpoints, limitations, or management.

When questioning our proper relationship with food, the paradoxical paradigm can help answer this question by immediately providing an equation for this classification, starting with its two bookends, and then deciding where the needle should rest. This most basic relationship we have with our environment, consuming food, can only be polarized into its two most basic, opposite purposes: living to eat or eating to live. Then, when the cross-examining process begins, it will be easy to see that either extreme end by itself can give us trouble. When people live to eat, their perception of this eating pleasure is out of proportion and out of control, leading to a chain reaction of problems caused by too much eating and too little emotional energy left for other important areas of life. In contrast, when you eat to live, it turns what ought to be a tempered, pleasurable occasion into a cold ordeal and an undesirable obligation, like taking medicine or vitamins, denying this necessary requirement the delight it should bring and preventing the satisfaction derived from enjoyable meals that enhance or shape our social affairs.

After properly setting up the equation for analysis, the logic will become obviously clear that a balance between these two extremes would be the most ideal and healthy arrangement. Eating good food with good people should be a great experience, but not too great. It must be kept in perspective to its overall effect on one's life. After the math comes the focus and energies needed to achieve and monitor this goal, a task that isn't easy because no acquired balance is easy; remember that balances are evenly matched struggles, not void or complacent. Then there is one last, main ingredient that adds to the success of this commitment, and that is the equation of thought—logic and emotion, previously analyzed—and the equation of emotions, love and hate, which is discussed in the next chapter. Therefore, when we are looking for the right perspective to have in any category, the quicker we reach and control a balance between the two detrimental extremes, the quicker we will have success. The easiest thing to do is be one-sided, and the next easiest thing, perhaps, is to have the right knowledge but not the control, which takes time, discipline, focus, and monitoring.

Having discovered in all these graphically demonstrated life experiences the exact spot to draw the line for success has been finally pinpointed in a very easy mathematical equation: right down the middle between two equal opposites, which are the same thing except in reverse, thereby and most importantly laying out the clear realization and legitimacy of both sides. How simple it is: no more guesswork, no more mysticism, no more blurring of the definitive lines. Hopefully, this gives confidence to anyone struggling for the answers to life's basic questions that the present age has blocked from us, confidently knowing that there is structure in the most important areas of our lives, an orderly system that we can learn from and manage well with its effective discerning tools on our own.

Maybe no one should tell this to the experts, but guess what? In spite of their denials, it turns out there is an acceptable way to find answers after all. A sound and valid process does exist for examining the main, broad, and simple areas of our lives, not just in math, chemistry, biology, physics, etc. These paradoxical exercises and examples have proven that we can apply math to all areas, whether they are everyday living, critical thinking, judgment, or moral values or even religious, social, human relations, art, and philosophical subjects. Finally, we can dispel the present and false notion that these subjects cannot be measured; as all these demonstrated continuums have shown, true science can be applied to all relevant or irrelevant aspects of life because of one simple, irrevocably logical equation: math is life and life is math.

7

Putting It All Together

I do not know if I totally agree with Socrates that "the unexamined life is not worth living," insinuating that a mentally dominant life is so much better than a physically dominant life that there would be no need to live if his priority had the reverse advantage. Choosing one slant over the other could be hard, although living one's life with an inquisitive and energetic mind in a lethargic body to support it may be better than living in an active, energetic body with a listless and aimless mind to guide it. Thinking and doing are extremely important activities, but, when they are done together with the same amount of intensity, life can be far more rewarding and proficient.

When these two cooperatives are functioning in a balanced agenda, another pair of coinciding opposites are needed as well: choices and information. We can only think and act within the bounds of the choices we have, and our choices are only as good as our information. It is great living in a time where we have so many options, but it is also dangerous and unfulfilling when there isn't enough correct information being given to help the decision-making process. There are two wrong alternatives: (1) too many choices without adequate information, and (2) adequate amounts of information but not many options to choose from. When our two opposite powers, brain and body, pair off evenly with two other fulfilled opposites, choices and information, the ingredients come together for the ability to do something great, as in good, or something equally bad, or both; the choice is ours.

When to Choose Both

The simple issue of making choices is no different than planning and constructing a building or a house. As expected, you start out by laying foundations and then proceed with the progressive levels of construction, finalizing with the finishing touches. Perhaps the most basic foundational equation—along with a sound mind and body, bountiful choices, and great information—that is extremely helpful in learning to be confident, qualified, independent thinkers worthy to control our own thoughts and destiny is the emotional one. There are many equations that come into play that successfully guide our lives and choices, but none is so important as the mental energies that help determine our wants and rejections or likes and dislikes. Many times, "The mind is willing but the flesh is weak." This is why we need to understand our emotions the best we can, because nothing influences our decisions more. And, of course, the best way to learn is to dissect this category into a paradoxical equation for a thorough cross-examination that provides proper labeling, a clear vision of how the components work together, and, ultimately, how to control them for utmost competency.

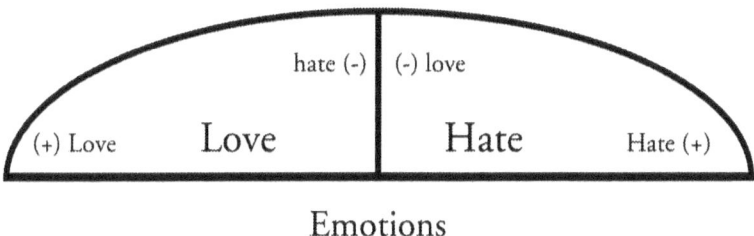

hate (-) (-) love

(+) Love   Love   Hate   Hate (+)

Emotions

I saved this important continuum for one of the last, as it is perhaps the most difficult for our advanced minds to accept because of hate's hidden value and agenda. Ironically, this is the emotion we love to hate. As much as I hate it too, it is no doubt part of the woven fabric of our thoughts. It is part of the necessary training ground for emotional associations and relationships that build from likes and dislikes, whether animate or inanimate, as well as being the requisite, attached, and equal partner in nature's dual roles. Can you imagine creating a hero in a play if there is no villain? Or how would you like being married to someone who never experienced jealousy, not even hating the thought of your leaving and loving someone else? Being nonchalant would mean that neither hate nor love was felt in his or her relationship with you. Therefore, sensitized feelings for or against would be nonexistent.

I wish we could experience love without hate, but that is the mental energy as an opposite motivator needed to anchor, magnetically cling or resist, offset, and distinguish the intangible, varied feelings of love. However, if two equal opposites are the requirements for balances and successes, for this is what these two are or ultimately should be, then what creative change should be made? Thank goodness we are not in charge, thereby concluding that, if we want to have love in our lives, we must also recognize the equal need for hate. Love and hate work together, as all coordinating pairs do, and, if this equation could be changed into just the single trait that we favor, making it different from all the other working and successful continuums, we could achieve everything but love. How loving would that be?

As mentioned before, if you don't much care if things are right, then you automatically don't much care if things are wrong. This is not a voluntary mental preference. This is an involuntary, mathematical, reactive, default system. If there is not much hate in our lives for anything, then it also means there is not much love in our lives either, whether we like it or not. Someone important in history once made a statement describing this emotionally depleted condition, declaring that the state of being lukewarm is not a good one to be in; it represents a stale and unfeeling mental condition. He suggested that it is better instead for a person to experience either extreme, knowing that one extreme automatically leads to the other, thus creating the possibility of both sides, love and hate, being activated and eventually becoming balanced.

Having strong unbalanced feelings is a start and is perhaps better than a very small balance of emotions or no feelings at all. I think it would be easier to redirect someone's strong feelings if they are misplaced than to try to motivate someone whose emotions are mostly dead or whose feelings were never cultivated in the first place. Learning how to hate the deed and love the person and how to practice a biblical scripture that admonishes us to "love the good and hate the evil" seems to be a far better trait than for a person to be emotionally removed, feeling no distress or responsibility toward others, for instance, having no compassionate need to help a poor victim if a brutal crime is being witnessed. Nor is it a good position to never experience being turned on or off emotionally to anything. This is the characteristic we associate with robots and fabricated outer space aliens, not humans. As the old saying goes, "It is better to have loved and lost than to have never loved at all." No doubt this human emotional experience enriches one's life.

Why is it that certain leaders from our past have been called great? Is it because they were simply mild mannered and extremely benevolent, never causing any harm or hating anyone? Of course not. For example, Alexander the Great and Peter the Great were called great because their cruel brutality toward others was equally matched by the tremendously good and great achievements they accomplished for their own country and people. Alfred the Great, "one of England's finest monarchs," was not only a successful warrior by killing invaders, but he made notable contributions to government, education, and literature and helped protect the average man from civil wrongdoing. They had both enormous hate and enormous love for others; they did enormously loving things and enormously hateful things. Their great destruction was matched by their great construction, the bad being equal to the good. Hitler was never given the title of great because he didn't accomplish anything for his people but pain, humiliation, death, and destruction. It was a total setback and equationally one-sided.

Using the general transfer of basic knowns to unknowns is just as effective and instrumental in the equation of love and hate as it was for the other obscure subjects. What most people can understand quite well is how budgeting their money works, as well as the need to budget their time. The same exact procedures and objectives that pertain to these categories can and should be applied to the emotional continuum. Experience and financial common sense alone would tell any budget keeper that the only way to balance a budget is to compare income against expenses, equally divide and allocate the appropriate amounts to a realistic list of needs and wants, fixed and flexible liabilities, plus compulsory and optional purchases; then you apply the discipline to carry it out. I am sure we've all been there at one time or another. The inexperienced money handler will usually take care of the immediate wants and desires and then pay the bills with what's left over, if there is any, getting away with it until, somewhere down the road, reality and creditors eventually pay a visit.

However, if the money is wisely budgeted by organizing priorities and determining the correct dollar amount to cover all expenses, with patience and time there will be financial success. A budget means you have only so much income to work with. If you put too much in one account, there won't be enough for some of the others. For instance, if you spent too much on entertainment this month, there might not be enough left for the rent. Just like all the other equations, it has to be organized in a good system and then monitored. The same transactions apply to time. We are only given so much time, and, if we don't consciously direct or manage it in some form of discipline, it will quickly and shockingly run out before we're done.

These same rules are pertinent to the emotional equation as well. When we fill up the love side in all the appropriate sections with the right priority amounts, making sure there is enough of this intangible energy left to stretch across all its territory and obligations, there will be peace of mind. Is it right to have such excessive love for self or spouse that there isn't enough left over for children, neighbors, pets, or the environment? Even with love, too much can be too much before it completes its destination, the whole spectrum; it's half of the continuum. The danger in concentrating a disproportionate and unmerited amount of love in one spot, such as one's self, leaving wide deficits elsewhere, is that the situation will automatically and explosively be equally reversed on its opposite side, hate. The exact same concentrated amount of energy on one side is duplicated on the other side, being automatically released and directed at the opposite subject, which would be others. Like the physical laws, the passionate forces are equal and opposite also. However, when the emotional energy of love is spread out into all its many areas in the proper priority amounts, it creates a stabling force, which, of course, is balance.

Conversely, the proper guidelines for hate's half of the equation are precisely like its opposite half, love. The energy force for this side must span a wide range of subjects as well. It must not use up all its abhorring energies in one or a few areas and not leave enough for all the other proper things to hate. How unjust it would be if we didn't have enough mental power to hate all that is bad or evil? Can you imagine the inhumanity of deploring stealing and murdering too much so that there wouldn't be enough emotion left to hate social injustice and depravities, child abuse, spousal or gender abuse, animal abuse, verbal abuse, environmental abuse, homelessness, cheating, gossiping, backbiting, bragging, arrogance, negativism, imbalances, etc.? In reverse, how unjustifiable it would be to loathe cruelty to animals so extremely that there is no anger left for cruelty to human beings or to hate a person so much that there's no more energy to contend with or perhaps recognize other enemies, thus creating an emotionally dysfunctional and vulnerable state of affairs.

Just as the ideal unity within a category is an equal blend of its opposites, the same case holds true when one continuum parallels and coincides perfectly with another continuum. Again, there are too many dangerously wrong associations and arrangements to mention, but the right combination is simple and easy to explain because it is only one. Accordingly, when they match up in the right order, cognitive success will be overwhelming. The perfect parallel to the love and hate equation is the continuum of good and bad, but with the right match up. Unfortunately, it is all too easy for human nature to love the bad and hate the good. To argue otherwise, one should be reminded why we must have mothers and fathers, teachers and bosses, policemen and soldiers in our lives or why nearly half of our population is obese, not to mention mankind being notorious for following the path of least resistance and doing what comes easily, not what is right. If our natural tendencies were reversed, none of us would so easily eat junk food the way we do, do nothing, be impatient, talk bad about others, be selfish, or get even.

The easy choices are usually not the best choices. To choose good over bad eventually requires a long stream of loving and not so loving disciplinarians and devoted teachers, culminating in self-conditioned, responsive controls and self-appointed checks and balances towards good behavioral choices at some point in our independence. Children and people left to their own devices and vices self-destruct. Unfortunately, there are not many of us that can totally live up to these two corresponding continuums, but the closer we come, the nearer is our journey to success. In conclusion, when all the emotional energies of love and hate are consumed in the appropriate places of the morality equation (good and bad) in the right amounts, having lined up correctly with this dual partner in its fullness, then loving the good and hating the bad will be an accomplished, highly rewarding climax.

When questioning other personal issues in our lives where answers are vague and unintelligible, the quicker you are able to set up an equation for cross-examination and evaluation, the sooner the answer will come and the sooner you can react appropriately. For example, how do you know sometimes whether you are being nosey or being genuinely interested in someone else? Instead of being complicated, it can be a very simple diagnosis, because there is a fine, comprehensive line between being interested in other people's lives, which is good, and getting into their private affairs, which is bad. There are two wrongs: (1) to be so uninterested in other people's lives that no inquiries are made or no thought is given to them whatsoever, and (2) to be too interested in other people's lives so that it goes beyond the point of being interested to becoming nosey and minding someone else's business. The checkpoint is very easy to detect: one side is unselfish and the other is selfish. If you are a very warm and caring person toward others, you will be interested in their lives up to the halfway point and nothing less, but, if it crosses the boundary line into the selfish and prying side of the equation, then obviously you have gone too far. You just need to ask yourself honestly which one you are being. If you become proficient in these kinds of checks and balances, you can quickly put things into the right perspective by pushing the needle slightly forward or backward before any harm is done.

There are so many other mental characteristics that can be evaluated this same exact way: for instance, guilt. Most people think guilt is bad and should be done away with it. However, it is both good and bad. Guilt creates a good conscience, but, when it has gone beyond its full potential to help us do the right thing, then it becomes destructive. The opposite half of that equation is guilt for things we can't do anything about or we had nothing to do with. It is fruitless and regressive to be guilty about something we can't change or have no control over. Therefore, there is a healthy and functional balance that can be achieved if one does not attempt to eliminate either side of this paradox. Our prisons are full of people whose consciences have been severely annihilated or were never developed in the first place. The bottom line is that it is mentally sound to have the equal capability of feeling bad about the things we are responsible for and not feeling bad when it is not our responsibility.

What about helping others? What could possibly be wrong with that? When set up in its proper continuum, the answer is quite simple. When helping goes beyond the point of being helpful, it has gone too far and becomes self-serving or controlling, which is a very selfish thing. The same is true with fear, which is good up to a certain point (halfway), such as respecting danger or authority. But, when fear becomes excessive or we're fearful of the things we shouldn't, it hinders us from being courageous when we should, thus keeping us from standing tall and doing the right thing. Anger fits right in with these personal categories, for it can be very constructive when channeled for the right causes, such as making necessary changes. However, when anger lingers beyond its usefulness, it just festers, having no place to go or any justification for being, and, when this energy is not redirected, it becomes explosive or, sadly, turns to lingering bitterness when prolonged in an unproductive state. On the other hand, can you imagine someone who never gets angrily upset, no matter what the circumstances are, who never fights for any reason or takes a stand on anything? Anger, therefore, should exist, for it can be both good and bad, depending on its cause and effect and point of balance on the scale of justice.

In other words, if we treated any personal classification like a good budget, a timely schedule, and good math, we would get the same beneficial results and correct answers. Financial responsibility and spending one's time wisely are, of course, no accident. It takes time, energy, thought, control, and a good system. I hope these many life-experiencing demonstrations prove what system that should be. The only method that assures an organized learning field and a controlled, personal environment is, naturally, the paradoxical system. Finally, and most importantly, managing our emotions prudently is also paramount and foundational to the success of the other categories in our lives, for it underlies and influences all our decisions. Not only do we need to lay out each individual equation (one) properly to enable efficient organization and management for our actions and for our thoughts, but we should also prioritize by lining up all the equations in our life in their best arrangement and support for each other through this exact same unique method in order to maximize our lives.

Prioritizing and balancing all the personal categories that make up our lives, fulfilling each account in a healthy required manner, is naturally of greater importance than properly arranging and managing physical things such as a software program, a filing system, or even a recipe. In any of these cases, when something is missing, deranged, unbalanced, or poorly defined, there is havoc to pay, in which the whole system can be threatened. A visual and simple sampling of this incompetence is a cake that is supposed to be light and fluffy but could fall and turn out flat if just one small ingredient was missing or there was too much of this or too little of that. Our physical maneuverings and ideas can usually survive a lot of botched attempts at organization and improper balancing acts or out-of-order priorities, but, for our personal issues, this may not be true.

The famous family counselor on television and household name, Dr. Phil, psychoanalyzes his clients in a way that usually surprises them. How many times have we watched him tell someone, for instance, who is suffering from a lack of control or too much control in a certain area of their lives, such as an eating disorder, that this problem has nothing to do with eating. Instead, he points to other equations in their lives that are way out of balance, such as emotions, self-evaluation, poor diet and living habits, or poor scheduling of their time that is causing their chronic eating syndrome. Then his prescription for their distorted, destructive, chaotic behavior is to clean up the other problem categories, which are the real culprits causing this unjustified control or lack of it instead of the act of eating. Once this is done, the disorder will get better or go away. Just like a budget, when needy accounts do not get their proper due and others are getting too much, the whole system can break down and cause bankruptcy.

I used this psychology, in reverse, years ago when raising my four children. I can think of two behavioral techniques that I purposely used to influence effectively and subtly several other categories in my children's lives that were more important than the ones in which these methods belonged. In an age when saying "yes, sir" and "no, sir," for example, was not mandatory or necessary, I decided to impose this on my children anyway, not because it was important for me or society, but because it could be important for them. In my estimation, it was a daily and subtle way for them to practice respect for their elders and those in authority, which, for a start, would help them have good relationships with their teachers at school, a foundation that was worth installing. My hope was that it would have rewarding and reverberating benefits for them throughout their lives, not just a personally gratifying response to me and my commands or questions. As a result, my children turned out to be sensitive and thoughtful adults, so, in theory, it must have worked.

The other valuable behavioral method I used, which strikes fear in parents today with the threat of being arrested or having their children taken away from them, is spanking. This brings in the old dreaded question: to spank or not to spank? It has gotten such bad coverage and is so one-sided these days that it is a subject hardly anyone discusses anymore for fear of retaliation. However, having had the experience of being a mom and now the joy of watching my children be parents, plus reading all the currently weird heresies about child rearing, which is sharply contrasted by my knowledge of the true meaning of success, I know without a doubt one very beneficial evaluation that could tremendously help today's parents is to present a balanced continuum of spanking. Like all the other categories, it can either be good or bad. When done either way, spanking can have resonating consequences that cross over into many horizons, affecting and covering many categories in a positive or negative way.

Therefore, I agree with Dr. Phil's methodical insight that it really has very little to do with the actual act of spanking a child and has everything to do with the general makeup and personal agenda of the parent that chooses to do or not do this type of punishment. This controversy, to spank or not to spank, could speak volumes for the good and the bad of this issue, because there are so many different individual parents with different mental agendas. It is just a tool that a parent can use for harm or for benefit, depending on his or her intentions and mannerisms. Just because you choose not to spank your children does not make you a good parent. Nor does spanking make you a bad parent. It is mostly who the parents are and the reasons behind what they do that matters. There are basically two wrongs: (1) having all the right reasons to spank a child from the person with the right position and integrity, being denied the freedom or justification to exercise this choice, if deemed best, and (2) having complete freedom, just because they are parents, to use spanking for any and all reasons, no matter the cause, consequences, intensity, or the integrity of the one administering the punishment.

Spanking children should have more to do with implanting wonderful attributes in children than the issue of the mere punishment, for, when it is done right, it can be a two-way street. A punishment that helps mold little ones into being better human beings creates an environment where there is less need for punishment, making parenting much easier. Any isolated idea away from its proper associations loses its correct meaning. That is why raising children is a very complicated task and should be so; being a parent is one of the most important jobs anyone will ever have. Therefore, it should be more complex than to just decide whether it is okay to spank or not. On the other hand, it is a tool that shouldn't be denied good parents when it works the best for their children, just as it shouldn't be an option for those wishing to do harm to their children or satisfy some distorted rage within their own corrupt being. Raising children and the methods to do so should have many good intentions and agendas, as the following list displays:

1. The goal should not just be to raise children, but to train and invest in future adults.

2. Children's attitudes are extremely important. How they are raised, along with their inherited genes and environment, determines their behavior. So getting a child to behave or not is almost immaterial compared to what he or she is learning towards being a good and responsible adult. Being still and quiet is good, but being respectful and considerate of others is better.

3. A positive and cooperative environment is like a chain reaction, for it creates many positives. If children are taught that the world evolves around them, what a rude awakening that is going to be when they eventually face the real world. On the other hand, if children are made to exercise unselfish and thoughtful deeds when they are little, doing it when they are big can come as second nature to them. A child who is never discouraged from or sharply reprimanded for being negative will find being positive almost impossible as an adult. Providing a positive environment by eliminating as much negativism as possible will go a long way towards a child's healthy outlook on life.

4. To have a balanced home environment, one must recognize there is good and bad about everything in our lives, and a good monitoring system should be in place. There should be work and play, children enjoying their childhood and children painfully preparing for responsible adulthood, rewards and punishments, praise and shame, individual interests and family interests, and the list could go on. Thus, success lies in the balance of all the above.

Consequently, when you have these ultimate goals in mind, it brings into better perspective the methods of child rearing: it gives you reasons why. It can help make a lot of the confusing garbage coming in from the media easily recognized for what it is when compared to a working and balanced agenda. When raising children to satisfy the public or for one's own selfish intentions, it will just make a bad situation worse. I doubt if any mature adult would disagree with the above optimistic goals. However, many will staunchly disagree with a method that could help make these goals a reality for parents—yes, spanking, in any shape or form. But this category is really no different than all the other personal equations that were analyzed, for it too can be both good and bad. Again, if you read current books on this subject by the majority of experts, you will only hear the bad side, making you feel like a cruel barbarian if you even think about striking your child, thereby confusing and thwarting the issue even more. I want to demonstrate, not only through my experience, but also through the logic of the paradoxical continuum, that there is a good side to spanking. This present-day controversy is in desperate need of being balanced.

I was not always successful, but most of the time I tried to relate my daily interactions as a mother to a long-range plan for my dependent children. This helped me cope with such immediate disappointments and frustrations as a child spilling a whole gallon of milk on a very small kitchen floor—yes, it happened. By immediately cross-referencing this situation in a broad picture spectrum in my head, it helped me react a little more calmly, knowing this was not intentional, no permanent damage was done, and, most importantly, I knew the way I reacted to this situation would help my child put this incident into the right perspective as well. Another tactic I consciously used was that, no matter what the occasion, whether it was an occasion that called for a spanking or not, I never refused a child a loving hug if that is what he or she wanted. The sticking point was that, even though the hugs were always available, it was not a license to do the "no-no" again. The repeated crime resulted in the same repeated punishment. I made sure the two things were always separate and consistent, giving him or her a clear vision that I always loved them, no matter what, but that a wrong will always be a wrong and will merit the same consequences each time. Confusion and inconsistencies are two of the worst enemies of child rearing. Last, but very important, I made a point not to show anger when rendering a constructive and positive spanking.

As the children got older, I was able to use spanking less. But, the reasons were many why it worked so well when they were very young. As opposed to the current method of long time-outs or the endless wrangling, negative arguments that are so common now between children and parents, I can give two good reasons why I preferred spanking over these modern conventional methods: (1) it took a situation quickly out of the negative into the positive, and (2) the reason it was usually quick was because I found that a perfect spanking, defined as bad enough to make them cry but not enough to cause any lingering harm, changed their defiant attitudes drastically into sweet ones. That was one of my main agendas. I felt that, if you were not working with attitudes, it was a lost cause. The attitude, not the punishment, is what they will carry with them into the next generation. Every time I tried sending them to their rooms or denied them certain privileges, it only made them more angry and unrepentant. Also, if I didn't spank them hard enough for them to cry, it wouldn't break their antagonistic spirit. Nothing, I mean nothing, sweetened them up like an appropriate spanking, as long as it was balanced with all the other personal agendas in their lives.

The child rearing experts, many of them having no children of their own, who strongly recommend that you never spank your children must ask themselves whether the children in this generation who don't get spanked are better behavior-wise than the previous generations who were. If we are going to base our opinion of successful child rearing on a condition that has to be free of hard times, physical punishments, and hard work, then our nation was shamefully and wrongly established from generations of immigrant families who did not know what arguing with parents was, who could count on being physically punished by loving parents when they did something wrong, who didn't know what temper tantrums were, who didn't have much, and who didn't get to play all the time or get into trouble because they had too much work to do. Can you imagine how it was possible that we could have such a great nation with the likes of this kind of savage parenting and upbringing? And how is it now? If you haven't visited a public school lately or the average home, you should do so.

I am not recommending that we go back to "the good ole days," but there is a balance, and the question should not be whether to spank or not, for spanking can be both good and bad, but, if done correctly, a parent should have the freedom and option to decide when it is a good time to spank and when it is not. If that is the tool and method that brings more positive results to all the important categories in a child's life, as it did for me, then that should be a parental choice, not a state or public choice. I hate to use the following old cliché, but I must: "You don't throw out the baby with the dirty bath water;" you keep the baby and get rid of the dirty water. Therefore, the main concern and emphasis should not be on the method, but on the results it produces one day: the overall package of a productive, positive, and mentally balanced adult.

A Place for Everything and Everything in Its Place

It is self-evident that we can make better choices or systematically arrange something better when all our options or objects are lined up, available for an overall examination. After that, the next best thing is to have is a good system for selecting and organizing all the items. I do not think anyone would disagree that the best organizing system there could possibly be is to have a place for everything and everything in its place, as discussed in an earlier chapter. With the general transfer of basic knowns to unknowns, I want to demonstrate that, if you know the ins and outs of how this popular and successfully organized method works, having applied and perfected it to some extent with material things, you can understand and transfer the same known basic principles and mechanics over into any obscure, personal, intangible, or complicated category and be guaranteed the same beneficial results. When you have a system, no matter what it is, that is the best it could possibly be, there's a reason for it. Believe it or not, there really are causes and effects, there really are questions and answers, and, systematically, there are definite reasons why.

It is a very simple arrangement that can be easily grasped, but carrying it out can be difficult. However, once you've mastered its systemizing techniques, your living habits conquered and refined under its commanding discipline, it is knowledge and mastery that you can take with you into other fields and subjects, and, if possible, you can systemize your entire life. Therefore, the following instructions and advice in organizing one's personal belongings, for example, are not just to help execute the procedures of this known organization, but to help this wonderful structure become the catalyst by which its transferable, basic principles can be applied in organizing vague and unknown categories or issues of greater importance as well. Consequently, the following guidelines are intended for organizing one's things, but even more so to show that, if the same rules of engagement are conceptualized and outlined, they can be practiced in all aspects of one's life. At the same time as you are learning how to best arrange a closet, you can also figure out, using the exact identical tools, how to best organize and control your thoughts under the same pattern of laws.

- Start small and start somewhere. The main thing is to get started, and the easiest way to be successful is to start with an easy category. Most victories come from making small, consistent steps over a long period of time, instead of occasional giant leaps.

- Little successes add up. It is important that at least the first attempt is relatively easy to succeed in, because success is contagiously encouraging and can add up to big successes, eventually laying a foundation of connected and organized categories that support each other.

- Set up a continuum. When selecting a category, you must reduce it to its lowest denominator and main purpose. Then, determine its two main opposite bookends. Success depends on these initial steps because laying the foundation has to be right on target. For example, the title "perfect organization" is the best possible label for a place for everything and everything in its place, because it cannot be reduced to any other defining purpose and no other category shares this same characterization. The next required step of dividing its two main opposites, their actions, and objectives will be relatively obvious because they are always the same thing except in reverse. Therefore, the two bookends for this category are, of course, creating the perfect place and system to store everything and the reverse follow-up of putting everything back in its place, the same as all and one, parts and whole, container and contained.

- Use the seesaw effect. This means working at both ends at the same time. Adjustments will need to be made along the way, because, in the numerous practices of storing and retrieving, each opposite transaction will reveal more information about the other until the final arrangement is perfected. If you set the arrangement of all things in stone before thoroughly examining and exercising the full characteristics of the objects to be stored or the integrity of the storage assembly, you may have to scratch it and start all over again. It needs to be a flexible learning and guiding situation along the way. The active process of storing and retrieving enhances the learning field, thereby calling for necessary adjustments. I call this back and forth action the seesaw effect because it is most beneficial when the motion is constant and fully extended, just like the playground ride.

- Variety adds spice. Just like the seemingly endless variations that exist between opposites within a continuum, offering a limitless learning field, the same situation holds true for the various equations in one's life. In studying the different categories, each will render a slightly new angle of working an equation that none of the others can within the same universal framework, of course. Each equation that is tackled and balanced adds a new and interesting twist to the system, giving new information; no two will have details that are exactly alike.

- Transfer the known to the unknown. When you are working a new category and you get stuck, go to a known subject that you understand well and transpose the basic fundamentals over to the new unit. If the example is a proper working equation, it will display all the right double-action, back and forth balanced principles that all paradoxical continuums have.

- Failures will happen. Failures are inevitable and should be expected, but, hopefully, they will not devastate you. However, it is what is done most of the time that adds up and mostly matters. Persevering, focusing, and continuously working the equations towards a harmonious balance will pay off in the long run. If you practice, conceptualize, and have confidence in these proven laws, the successes will eventually overtake the failures.

- Cultivate endurance. Enduring something difficult can be too much to bear if we don't have a visionary goal, occasional rewards, and a certain time period to aim for. There is a wide consensus and rule of thumb that says it takes approximately three weeks to create a new habit. I believe it because it has worked for me as well. Having this relatively short time range in mind will help considerably when the going gets tough and the temptation to give up comes along. Keeping your mind focused on the goal, the correct road map, and checking off the time as you progress will help pave the way.

With balanced equations of mind and body, choices and information, love and hate, and the working knowledge of organizing a place for everything and everything in its place, whether physical or mental, the ability to make good choices will become easier in time and with practice. Once a good foundation is laid, the rest of the development will fall in place. The support and the practice of these continuum laws enables a mastery of discipline and skills that will, in fact, help with any future decision-making process.

When faced with a simple decision or a very complicated one, an experienced paradoxical thinker's first reaction and order of business will automatically and immediately be to divide and conquer with the first initial law of assessment: is it an either/or decision, or can neither side of the equation be eliminated? If both sides are valid, as all the equations illustrated above have been, then the correct thing to do is to choose both and proceed with balancing the equation.

When to Choose One

On the other hand, if the end result has to be one or the other, then the only option is to choose one of them. The following exercise is different than the other mental categories because, in order to choose something, you are eliminating something else. All the other equations were not either/or situations because both sides were essential to the unit and the equal balance of the two was the aim, but this is not so with the following equation. This continuum is the deciding platform witnessed in the courtroom, in the ballot box, and in the case of a spouse when you must choose just one. But, as demonstrated in the following equation, there is a reason for this as well.

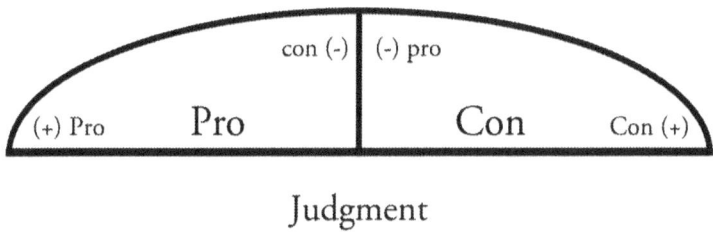

Judgment

This continuum is different than the ones above, because, when it comes to making definitive choices, such as pro or con, yes or no, a decision has to be made for or against, which is the same basic action except in reverse. Since all things are structured in continuum form, you cannot be for something without being against its opposite opponent. Consequently, when you reduce any category down to its most basic elements or stance, it will always line up between two opposites; therefore, if one opposite is yes, then its counterpart half is no. We make all kinds of choices all the time, from equations where we should choose both sides to occasions where we must choose one side. In contrast to incorporating both sides of an equation, this kind of decision-making calls for a situation when a compromise or middle status quo is not the ideal state.

Just the opposite of the other demonstrated equations, the middle line is not the place to end up in this type of transaction. Walking on the fence or line and straddled between opinions is not a favorable stance. Everything can be good and bad according to different circumstances, and, in this category, compromising is not the better choice because a decision for one side or the other needs to be made. In the other categories, compromise was the perfect concession for meeting each other halfway combining both complete opposing sides at the middle. However, even though this equation singles out one side or the other in the end, its initial basic structure and setup is exactly the same as the rest, having all the qualifications of a two-sided paradoxical continuum. This is what enables the decision process to be successful.

Perhaps the most valuable service the paradox has to offer is in the arena of judgment. This is where it really pays off. Measuring and monitoring the other categories is very helpful, but this is the one that indelibly determines our lives the most. Being able to make right decisions or the kind we want is the "mother" of all successes. Decisions we make on a constant, daily basis are always up for choice, but most of them are subtle and already slanted in the direction that was perhaps imposed upon us or that we chose early in life. Then there are the conscious choices we make as adults, according to the way we have interpreted life or happiness and the criteria or conditional trappings we are consumed in. Most of us want to make right choices, but we feel entangled in a medley of pushes and pulls in varied directions, or we are victims of misinformation. Therefore, the question is not only what are the right choices, but also how do we make them and keep them?

Ironically, the answer is a simple one, for there is a standard skeleton key to unlocking the mysteries of what is unknown. Mechanically, it is the same revealing universal procedure found in a paradox, but it can be demonstrated in a simpler form. As I have mentioned several times, the easiest way to understand the unknown is to go to a category that is known well and then generalize by transferring the basic characteristics and intrinsic transactions of that known category over to the one that is unknown. Consequently, in the realm of decision making, the best arena to go to for observation is in the courtroom. First of all and basically, a court case is divided into two opposite sides, plaintiff and defendant, neither side compromising—each being exclusively and staunchly against the other. As a result, this provides a full representation for both cases. Justice would not prevail if one side were weak or compromised. Next, the energetic, back and forth cross-examinations that follow create an environment for enlightenment and discovery. Thus, after all the facts are gathered and deliberated, the judge or jury weighs one side's evidence against the other in order to get the complete picture and try to make the right decision.

Everyone seems to accept the order of the court, a divided forum made up of two opposing sides, symbolized by a pair of scales, but many fail to carry that same simple order and justice over to the political system. Not being able to recognize the valid paradoxical format instituted in the American political platform, which has produced the greatest democratic government so far in the history of mankind, many want to do away with the two-party system. Systematically, the very thing that gives fairness to the courts also gives a healthy endurance, full representation, and constant checks and balances to government as well. Ironically, the very thing that causes a close and almost indecisive race, resulting from a perfect match up of candidates, is also what inevitably starts the rumors to change our system of electing a president. Some cannot see perfection even when it is staring them in the face.

On a personal level, this is the exact same process needed in making any decision or judgment, large or small. It is also the exact same thing as a perfect paradox: a complete division of two equal opposites, total representation and thorough cross-referencing. Only then should you have the freedom to choose rightly and let the chips fall where they may. There are two basic wrongs: (1) to make a decision by choosing one side before collecting all the information from the other side, and (2) to make no decision after all the facts are in from both oppositions. No one likes a hung jury. In spite of the fact, these two actions are more common than most people think.

As this study has consistently proven, there are two sides to everything, and they are opposites. How many times have we all made decisions based solely on what we perceived from the good side and totally ignored the possibility of there even being a bad side. Getting married and getting credit are two prime examples. How many make a thorough examination of the unfavorable side of those categories before taking the plunge? I would venture to say not many. Great desires and needs systematically employ great emotions. We gravitate fully and naturally to the side we like, without pushing the spectrum all the way over to the opposite end, making a logical investigation of the side we don't like or want to look at.

Again, there can be two extreme wrongs involving marriage: (1) going into this union with strictly rose-colored glasses and unreal expectations, or (2) after realistically envisioning the equally bad side it causes excessive fear, rejection, or indecision. The reality is that married life is going to be both good and bad, but, like life and art, if your ideas and dreams are based on reality in the beginning, you can help shape and influence what your life is going to be like. However, if it's broadly based on unreality, the relationship may not survive the distortion and unmet expectations, resulting in extreme disappointment and divorce.

In summary, decisions are hard but necessary. In fact, I have found that the harder the decision, the more meaningful is the dilemma, and, in reverse, the easier the decision, the more shallow and meaningless is the subject. Or, if an important issue was decided too quickly or easily, it should trigger warning signs that the investigative equation is not equally balanced; more work needs to be done. The best decisions come from the soul searching of having to pick between two equally good scenarios, or vice versa, two equally bad scenarios or from wanting something so bad or it is so important that you accept the unfavorably attached consequences that automatically comes with it with your eyes wide open, rendering the cost worth it. The energetic, cross-examining exercise required in such a paradoxically balanced case makes it hard to decide, but, at the same time, that is what makes it a good decision; therefore, the greater the dilemma, and the greater the knowledge, the greater is the choice.

The continuum templates in this chapter have taught us the ways of nature and of mankind: how we think, how we function, and how we live. Just as the old saying goes, "We are what we eat," it is equally true that we are what we think, and, therefore, we live according to what we have learned. We have also found that the true learning situation includes both opposite sides of a continuum, in their completeness, including the unfavorable ones that are so distressing and hard to accept yet even harder to deny. They are an equal part of the design, somewhat as a necessary evil, to offset, clarify, and cultivate as well as motivate the good side. Then, after we are well educated in this paradoxical process, it is almost like poetic justice that we should arrive at the following conclusion, which is ironically appropriate, as to how to define the two opposite bookends for one of our main clarifying or discerning paradoxical tools—the decision-making continuum—the main issue and dilemma we started with.

Now we can safely come full circle because the polar ends that fit this categorical continuum are none other than a paradox and a contradiction; they are undeniably like all the other equivalent opponents in an equation, the same thing except in reverse. Even though the contradictory side still requires laying out the facts in a paradoxical form for analysis, only one can be selected. For example, in the justice system, a person is either found guilty or not guilty in a courtroom decision, not both. Only one can be chosen.

Therefore, the two opposites to be set up in this continuum are (1) a paradox where two opposing sides are valid and both should be chosen, and (2) a contradiction where there are two opposing sides, but only one should be chosen. As usual, opposites always work in reverse. On one side, what starts out as one becomes two, and on the other, what starts out as two becomes one. Just do the math: on one side of the equation, you add, on the other, you subtract. Then, in the final analysis, it makes it easy to calculate whether something is a paradox or a contradiction.

8

Summing It Up

Hopefully, these demonstrated truths can help many to answer for themselves all the basic questions brought forth in the beginning of the book.

- There is a better mindset than to, satisfactorily, not know the foundational truths about the most meaningful things in life yet eagerly stockpile an endless array of facts and trivia that have no basic fundamental connections whatsoever.

- There is a formula that precisely measures all things in the same simple way in which we know with certainty that two plus two equals four and that logical thinking is logical regardless of where it is applied.

- There are basic fundamental principles that all knowledge and the pursuit of it are based on: one set of fundamental laws that apply, coherently and systematically, to all learning situations and all ideas.

- There are consistent laws that apply down to the basic core of nature that provide an automatic swing, which can take us on an educational journey abroad towards all categories, making the unique connection that exists between basic and general, simple and complex, as well as one and all.

- There are proper mechanical and fixed laws for the learning process and our understanding, procedures that are user-friendly and not just for the privileged few, only those endowed with genius-like capabilities.

- There is a course and natural guidelines that show us exactly how to go about defining, labeling, and governing our thought processes. It is a simple and specific measuring tool that anyone can use which enables us to measure truth and validity in all matters in the same precise way that we solve simple math problems.

- There is a scientific mathematical way other than "It's my word against yours" by which we can prove that what we consider to be true, which others may consider false, is really correct.

- There really is a process by which we can legitimately and confidently answer these difficult questions, thereby knowing with certainty what truth is.

In addition to these highly welcome resolutions of basic knowledge and fundamentals, questioned in the beginning and then enlightened through thorough and proper paradoxical examinations, there is another inquiry that needs to be addressed. Are we going to step up to the plate like the American colonists did and fight back for our individual sovereignty and intelligence? Are we going to continue to bow down to the kings and let the experts tell us what we should think? Are we going to turn to Dr. Phil to answer all our problems, or are we going to be a Dr. Phil? Now that we have found a viable way to stand up to these mental giants, judging and analyzing for ourselves what is good information or what is not, what is right for us and what is wrong for us, what is false and what is true, what is bad and what is good—then we can at last be sovereign, trustworthy, and responsible to find and stand up for our own decisions.

Knowing how to analyze anything correctly in its proper, natural setting, it is "we, the people," who need to learn how to discern wisely between all these choices and make good judgments. As mentioned before, this is something we must and should do for ourselves and not leave to others. We are the ones responsible and best positioned to take the facts from the experts and custom tailor them to our own wants and needs. With practice and correct knowledge, we can eventually trust ourselves to become good and reasonable judges for our lives. With the right discerning tools, we can sort out what the experts have accomplished with their expertise, applying it to our lives or throwing it out. Then these essential guideposts that the educated hierarchy is ignoring or destroying can miraculously help us in answering our most basic and extremely important questions about life.

Finally, the Answer

This unprecedented age has generated a barrage of questions that has bombarded our society like none other, and the very fabric of our being—our belief systems, our constitutions, and even our reasons for existence—are now up for grabs. The average person is beginning to find it very difficult these days to defend and find reasons for even the most common things in his or her life. All of the old standbys are losing their grip on society: the little communities that shaped and monitored social behavior; the neighborhood churches and doctrines that watched over their flocks; the schools that entrusted the finished product with prepared-ness, brightness, and foundational loyalties; and the homes that ensured nurtur-ing, discipline, togetherness, and protection. Instead, the trend is now towards loosening up the reigns that bind these institutions together in order to provide more freedom, synchronization, and more choices. In homes, where individuality is shaped the most, there is such little order and restraint or guidance compared to a hundred years ago that it is almost to the point where anything goes and everybody is on his or her own to do and think as he or she pleases.

This current situation could sound very gloomy from a normal, one-sided view-point, but the good news is that, if you have your bifocal lenses on, you can immedi-ately see it as it really is, both good and bad, without having to rally a whole chain of disagreeing experts to either prod you into accepting their biased opinions or to be politically correct and have no opinion. Fortunately, you have it within your own powers to analyze this situation properly and see it in the right perspective all on your own. You can look at these unraveling circumstances and call the situation what it really is—bad—not trying to camouflage it with rose-colored glasses or do away with it, but setting it up in its proper equation and then building on its oppo-site half, which is good, and as a result getting a broad, balanced, and clear picture. The bad side is a motivating factor, anchor, and magnetic resistor to work from. Without this vital half of the mental equation, it would be like trying allegorically to maneuver in a gravity-free environment physically.

The dynamics of the good side of this modern-day dilemma is that it provides the freedom to exercise choices. The loosening up of the ropes naturally works both ways. If there is a better way of thinking and doing things, we may now have the freedom and option to implement them. When everyone had to think and act one particular, dictatorial way, it could have been dangerous to do other-wise. Therefore, the best option is to try to make the bad side work on behalf of the good side. Left to itself, the bad side is destructive. Without its opposite com-ponent, its reason for existing becomes pointless. Of course, the same holds true for the good side. Not only are both sides needed, but also the balance of both sides gives the best results. Too much of the good side becomes just as destructive or regressive as too much of the bad side.

Anyone studying history has to be struck by some of the really good things that have resulted from some very bad things. This is not an attempt to try to belittle or soften the bad, not in the least. Nor does working on the good side of the equation lessen the bad or take it away, but it balances it instead, just as a powerful and muscled body offsets the gravitational pull, the gravity is there nonetheless. Without referencing a whole lot of examples, I can think of one that is perhaps one of the most ironic. Recently, I discovered that historians are crediting the salvation and restoration of the English language, which is becoming today's world language in communications, to the grotesque demise of the Black Death, caused by the invasion of rats. During the time of William the Conqueror, the English language went underground, being repressed towards the rural areas of Great Britain, and was on its way out while all the lords and clerics forcefully spread French and Latin throughout the land. When the plague broke out from the rats' disease-carrying fleas, it affected mostly the inner cities and circles of authority, which spoke French. The scattered and rural areas were the least affected, not only freeing up the English people, but the English language as well. Otherwise, most likely, those of us speaking English today would probably be speaking French instead. Thus, enemies and heroes can come in all disguises, depending on your viewpoint. When you understand that there is a mathematical absolute that blessings come from adversities and adversities come from blessings, there is better structure, wisdom, balance, and control in our thinking plus fewer surprises.

Once these mathematical exercises are put into practice and become second nature, you will never be held hostage again. You will have the power to think, clearly analyze, and be allowed to have real answers. Real answers are not more questions, and real truths are not fragments of a whole. You must not pick and choose to observe and analyze only one side; you have to accept the reality of the whole divided picture. The only way you can do that is to set up all the facts, whatever the subject, in a whole continuum that contains all, leaving nothing out, in a cross-referencing spectrum of two opposing sides. If you do this, you will find yourself saying yes and no to a lot of things. It will be so much easier to answer those despairing and isolated questions such as "Where's it at?"—instead of wondering whether it should be family, me, career, home, religion, money or whether it is good or bad. You will automatically know that it is all the above. The answer comes in the personal satisfaction and chain reactive successes of having it all properly arranged and prioritized in a paradoxical equation.

I recently saw a debate on television of a popular evangelist who preached hope and positive thinking with a prominent news commentator who was paying the retaliating price for being honest and straightforward in his approach to world events. He questioned the evangelist as to how he personally could approach his real, "get down and get dirty" world of politics with the philosophy of looking at the bad and ugly with rose-colored glasses. The preacher assured the newsman that he was sure he was doing what was right in dealing with this harsh world, but he just reiterated his position that hope and positive thinking lead to righteousness, never offering the poor gentleman any real vision or solution as to how he could possibly transfer these rosy thoughts of good, hope, and optimism into his real, pessimistic, and ugly world.

If this preacher understood the true laws of nature and the laws of the paradox, he could have told this news commentator: "Yes, you fill a very needed position in our society, someone who stands up and tells it the way he sees it, but, yes, you are going to pay the price. However, when you pay the price, you will be equally rewarded elsewhere, because suffering brings happiness just as assuredly as pleasures bring distress." He could have opened up his mind to the real world of good and bad, positive and negative, but, the way it was, each was in his own fragmented, one-sided viewpoint of the world. One wore rose-colored glasses that hid the ever-real bad elements in life almost deceptively, pretending they aren't there, and the other had eyes of vengeance, with no vision of the subtle, just as real, good and positive characteristics that come from standing up for what one considers right. Neither one was able to crossover into the other's vision of the world. To have been paradoxically on target, one needed a more positive viewpoint to offset the overextended negativism, and the other needed the negative to hold back the overextended, unrealistic positive outlook in order to balance each one's personal equation.

This is sad, because the real answer is that both are right, but not separately. If one of the most fundamentally basic transactions in nature, magnetism, cannot properly function or exist without matching positive and negative charges, what makes us think we can? Therefore, putting aside the normal way of thinking and attempting to employ this new elevated dual concept, how do we stand up boldly for what we believe and take the retaliating consequences but, at the same time, feel good or positive about it? There is only one answer, and that answer is, of course, the automatic, instant, dual viewpoint you get from the paradoxical continuum. It allows you to see the big and combined picture of how the two sides, the good of it and the bad of it, work towards our enrichment and our enlightenment, that, without both, we would not have an advanced, intellectual perspective that comes with a full and balanced education.

If this preacher had known the marvelous inner workings of the paradox and been forced to live in this man's very unpleasant world for a while, he would have been in a much better position to assure him that what he is experiencing, along with his position, has its valued and proper place in the overall spectrum of life and should yield an equally countering, positive, and hopeful outlook or result. When we are experiencing pain, we can't feel good at the time because it hurts, but we can have the positive confidence during the crisis or unpleasantness to know that, when the pain is over, the pleasure will automatically return. But to call something what it is not or pretend it doesn't exist is deceptive and unproductive.

All these many little balancing acts in our lives not only help us understand and deal with each individual (one) dilemma, itself, but they should also give us confidence to know that all of life is one big design that is in the process of balancing itself as well. If all is in one and one is in all, in the end, when all is complete, it will have followed the same path, order, and design of all its little DNA prototypes (one), but on a much grander scale, thus leaving us with the conclusion that each little part represents the whole just as the whole envelops all its parts.

Consequently, the answer is that it is okay to experience both sides of an equation; it is okay to be angry up to a certain point, to feel guilty up to a certain point, to experience fear up to a certain point, to hate up to a certain point, etc. But some will say that it must then be okay to be bad up to a certain point. Well, let's see if what works in all the other equations works in this one also. If we put it into an equation of two opposites, one would have to be something like good bad and bad bad, or minor bad and major bad. Therefore, in the comparison of the two, that would have to make one good or preferable and the other one not, regardless of what you call it.

Also, if someone were to argue that in order to develop the love side we would have to work on being bad, I am afraid bad has already happened. It is the anchor or gravity from which we struggle to resist. If you don't believe this, why is it we have to force ourselves to get off the comfortable couch and exercise, eat salads, fruits, and vegetables instead of cookies, cakes, and pies, zip our mouths when we hear some really juicy gossip about someone we don't like, give rather than take, and think of others instead of ourselves? In other words, it is easy to be bad. It is the other side we need to work on.

The opposite situation is true as well. It is okay to love, give, share, and work up to a certain point, but all these good things begin to turn against themselves from that point on, not only in the intangibility of thoughts but in the actual tangibility of mathematical transactions also. That point is the halfway mark in the equation of a paradoxical continuum. Once it moves across the equal line of division that separates the two opposites, it begins working in the reverse mode, on the oppositional side, just as the blooming flower that crosses that same line. In overview, not only is it okay but it is imperative that all basic elements are included in our lives in a guided and balanced perspective. And the icing on the cake is that it shouldn't be guesswork where that certain point is. It is quite simple and precise in its measurement, as are all basic mathematical equations. Where is the confusion in knowing where to draw the line in the following exercise: $\frac{1}{2} + \frac{1}{2} = 1$?

Who Is Right: Socrates or the Experts?

Now, having gathered and analyzed a considerable amount of information, we are perhaps better equipped to answer the philosophical question: "Who is right: Socrates or the experts?" Are there answers, or are there just questions? The world of education not only tells us that it is absurd to try to understand the most basic, fundamental questions about life, but it is also politically incorrect to divide and label sensitive issues sharply—sensitive meaning all the many subjects that are personal and important. I can't think of any I can exclude, since it all connects and all things play a vital role in our overall makeup. Their main policies are to not draw lines, avoid pointing out differences, don't make waves or offend, and, by all means, never make absolutes about anything, leading us to the polite condition of having no preferences or choices, never taking a stand on anything, nor have strong feelings of love or hate, for or against; they tell us to just ride the waves, go with the flow and, theoretically, be an object not a person. To them, this is okay because answers aren't important anyway, just the questions and, of course, being safe.

On the other hand, Socrates was the first to use the analytical method of reasoning, a method that searches for clear definitions of basic ideas such as knowledge, virtue, and justice. His motto was "know thyself," not obscure thyself. Can you imagine the ineptness, if all the demonstrated equations in the studies that this book has covered had not been analyzed, dissected, and a line drawn between each one's two main components and if there had been no cross-referencing of automatic, contrasting, and interactive laws to observe? Yes, by taking a stand for what he believed, Socrates eventually became a threat to the status quo and lost his life for it, but he gained a significant place in history that few have rivaled. Had he been squeamish and kept his mouth shut or hadn't cultivated such controversial and astounding insights, he would have eventually died a few years later of old age anyway, having achieved nothing but a safe, politically correct position among the common masses throughout the ages.

The following quotation comes from *The New Book of Knowledge Encyclopedia* (pp. 191–192), rendering a brief history of philosophy, helping to explain the evolving process that has branched out and stemmed from eons of intellectual pursuits, and culminating into the generalized, scientific community we have today:

> Philosophy means a highly disciplined and reasonable method of criticizing fundamental beliefs to make them more clear and reliable. This method was first developed by the ancient Greeks in the sixth century BC. Thales, Anaximander, Anaximenes, Pythagoras, and other learned men began to speculate about the underlying causes of natural phenomena like birth and death, rainfall and drought, the perfectly regular motions of the planets, etc....Socrates (470?–399 BC) turned attention away from nature toward man and society. His teaching inspired his disciple Plato (427?–347 BC) to write his (Socrates) dialogues, the first masterpieces of philosophical writing. Their influence on Western civilization has been second only to that of the Bible.
>
> Plato founded a school known as the Academy. Its most gifted pupil was Aristotle (384–322 BC). Aristotle combined a concern for pre-Socratic speculations about nature with the Socratic concern for social and moral problems. He continued the task begun by Plato of organizing all the fields of human knowledge into a unified view of nature and man. This is called synthetic philosophy. Aristotle set down, in a clear and systematic form, the general principles of most of the sciences and humanities. He established logic, metaphysics (the study of the nature of reality), and art criticism as fields of rational study.
>
> Plato believed that ideas were more real than things. He gave mankind a vision of two worlds—a world of unchanging ideas and a world of changing physical objects. Aristotle did not believe in a separate world of ideas. He provided a vision of nature as a single system of things that can be classified (divided) by genus and species. He said that each natural object contains its destiny within it, as an acorn contains a tendency to grow into an oak tree. Saint Augustine fashioned a view of life as a stage on which the creatures of God act out a drama of good and evil. Aquinas combined the thought of Augustine and Aristotle. His philosophy later became the official doctrine of the Roman Catholic Church.
>
> Hobbes and Spinoza developed a mechanistic vision of the world in which all events are governed by strict mathematical laws. They said that with sufficient scientific knowledge events would be as predictable as clockwork. Descartes set God and the human mind apart from the world machine and created the view known as Cartesian dualism. Kant deepened the division between mind and matter by separating moral laws from laws of science.
>
> At one time all fields of study were accepted as parts of philosophy. Religion and science were particularly important in every philosophic system. But with the rapid advance of knowledge the sciences and the humanities separated from philosophy....The synthetic type of philosophy that organizes knowledge into a single picture of the world has become more difficult to carry out as human knowledge has grown in scope and detail....Philosophy

today is analytical in the Socratic style, rather than synthetic in the Aristotelian manner.

Down through the ages, the number one pattern and objective goal for these truth seekers was without a doubt the discovery of the ever-elusive universal thread (one) that weaves everything (all) together, or the basic foundational blueprint (one) upon which everything (all) stands. The main pattern and analysis abstracted from this summary is their common quest to find how that unification is set up, with disagreeing arguments about what is united and what is divided, which is dominant and which is recessive, but still with the same goal in mind: universal design. As we have the privilege and hindsight of looking back over these vast years of struggling experiences and progressive levels of human thought, it is somewhat easier for us in this day and age to bring all their suppositions together with an overview and better perspective or vantage point that a bigger picture has to offer, providing the ability of fitting them into a general and basic evolutionary evaluation (one). Without giving us too much credit, though, we must humbly realize that providing criticism or critical analysis is less innovating and far simpler than establishing the initial creative thought.

This universal aim continued down to the twentieth century with Albert Einstein, for example. He tried desperately to consolidate the laws of physics, specifically gravity and electromagnetism, into one unified pattern of behavior, but he failed in his attempts. However, it was about this time that the seeds of defeat towards unification began to set in. Gains to find the universal paradigm by great minds, if any, were soon overwhelmed and overtaken by the cold, not so philosophical advances of technology. Power and might had become right. In contrast, a safe, compatible, less controversial, material world soon reigned supreme, bringing with it the unquestioning and stifling policy of political correctness that protected it. Not only was it more exciting to have powerful gadgets and modes of transportation than to acquire honor and a belief system, but now it was the proper thing to do, giving us permission to live for today, forgetting about tomorrow, and certainly not wondering what all this means and where it is taking us.

Philosophy and moral causes on national and authoritative levels were pushed back to a semi-fictitious status, both justifiably and unjustifiably. Consequently, its importance waned considerably for two main reasons: (1) physical science and technology were suddenly and momentarily giving us a better world, with the public loving every beneficial advancement it brought forth; and (2) the field of philosophy did not make the same factual progress that the explosive technological revolution was making and was soon left behind, not being able to keep up with the real, current world. As a result of this isolation and with time, many philosophical theories became quite bizarre and unfathomable for the majority of us who couldn't or wouldn't follow an endless sophisticated maze of unimaginable speculations that only the author understood or not. If you don't believe it, just go to some of the Internet Web sites on philosophy and try to read some of the material.

When a theory, or anything for that matter, goes beyond the point of any semblance of reality or connectivity, it naturally loses its ability to exist or be defensible. This is basically why the study of philosophy is limited to a very narrow branch of the scientific field, and, for some experts, even that is a bit of a stretch. However, the good side of the issue is that this quandary is a natural developmental process in the community of paradoxes. Different categories have different rates of growth. For instance, it is best that the union between man and woman be deferred until both have had time and experience to grow into two mature, independent adults before trying to work together as a group. One of the situations that is so problematic in marriages is that some adults are married to children, some literally and some figuratively. Needless to say, this is not good, nor is it necessary to mention all the different warped and unproductive associations that form such unions.

Some subjects are best separated for long periods of time, for example, church and state. However, it is possible for these two separate entities to mix compatibly when a simple, unilateral governing system governs a people of like mind and religion. But the United States' unprecedented constitutional stance provides for a wide variety of freedoms, of which religion is prime. The best-case scenario, for the moment, until something better comes along, is to keep the present multireligious sectors all contradicting each other by each claiming to be the one and only, separate from the task of governing. If church and state were to mingle, which one or how many should be chosen? And, if one were to be picked, we would have to do away with the Constitution because freedom to take part in government would be denied all religions but one. On the other hand, if all faiths were represented in state affairs, which are many, it would be larger than the government itself, not to mention the religious wars that would ensue, preventing the main reason for its existence—the business of governing.

The same scenario holds true for philosophical concepts. There are and have been so many varied, conflicting, fragmented, and simply wrong theories that I believe we've been better off separating them from science until the right perceptions come along or until the right perceptions that may have come along are finally recognized. From another viewpoint, it is actually good that philosophy and physical science eventually came to a point of separation and sort of went their own ways for a while, because the general public can usually handle only one issue at a time anyway; the very reason the duality viewpoint is hard for most to digest, making an either/or, for/against decision much easier. Furthermore, I'm sure being able to take on the industrial revolution free of conscience or deep thought helped tremendously to bring about the present day proliferation of technical knowledge. Just look at history and some other parts of the present globe. Where shortsighted, restrictive religions and philosophies or superstitions are dominant over secular, there is very little industrial development. One opposite can inhibit the other's growth if it is not a good match; not only must advanced thought equal advanced technology for balance and harmony to exist, but, when it does not, it can be dangerous. Case in point, the Crusaders from the twelfth century AD were staunch Christian servants but obviously didn't entertain the thought that killing their converts was not good stewardship. Needless to say, the concept of loving one's neighbor, under the command of an army, did not take hold at that time.

On the other hand and back to the future, in addressing the ever-abiding bad side, how long or how far can a runaway, guiltless, thoughtless, but technically powerful society go until it becomes extremely unbalanced and self-destructs? Is there enough time to switch gears safely, bringing to the forefront and to full fruition the opposite side of the equation, which is moral obligations plus rationale, before it runs amuck? Putting mind back into matter, with the correct matching mindset, that is, would create a continuum of philosophy (thoughts) and science (technology) by allowing the possibility of sound reasoning and moral consciousness to catch up, thus producing a more safe and equal existence. The question is this: What is that mindset? There has been such a vast array of differing philosophical ideas over the years, some good and some bad, that it can be confusing to know which one is right. On the other hand, insightful and well-organized concepts, on the human level, may have already reached their pinnacle peak during Greece's golden age.

The three masters, Socrates, Plato, and Aristotle, were mental giants, but they were still in the dark about a lot of things. Unfortunately for them, they were missing an enormous amount of physical knowledge we have at our disposal today because of all the technology gained from accumulated facts and the immensely high-tech instruments available to us,—but, when it comes to logical reasoning alone, they may have come closer to the truth than any other great philosophical minds. Paradoxically, not only could they have used the current, tremendous source of scientific information in their day to verify their logical dissertations, but, in reverse, maybe the present educated community of today needs to catch up to their incredible mental feats of insight and genius generalizations (making sense of phenomena or connecting the dots), as well, such as "each natural object contains its destiny within it, as an acorn grows into an oak tree." Isn't this just another way of saying all in one and one in all—all acorns grow in an oak tree (one) and an oak tree (its one DNA) is in all acorns? Even without knowing the existing knowledge or technicalities of DNA that can now be detected through complex and sensitive instruments, Aristotle rationally concluded that the creative information that made possible an entity in nature to redesign and recreate itself in the exact and complete form must be inherently within, like an embedded seed.

To sum it up, if these phenomenal philosophers could have had our sophisticated information and we could have their deductive powers, what a grand match that would be. Nevertheless, even if Aristotle could see this universal order clearly in his head, without the assistance of molecular proof, spreading the idea out to the masses was an entirely different matter. I am sure the limited knowledge about the physical environment that engulfed the people of that day inhibited their capability of understanding the logic and soundness of this conceptually thought-provoking blueprint that this great thinker surmised so very long ago. Again, opposites are our discerning tools; to know one grand structure, you must know the other. The more we are capable of understanding the complex physical transactions that produce the world we live in, at the same time, the better we can understand the intricate mental transactions that functioned in Aristotle's complicated head.

There may have been others from the past or some living today that have shared the same concepts of unity that were either skipped over or kept dormant. In my recollection, there is only one other exception that comes to mind of someone who has demonstrated the understanding of this paradoxical insight, again, on the human plane. Solomon, the once great and powerful king of Israel, was credited in the Bible as having wisdom at one point in his life that exceeded all men before him and after him. Until I discovered the universal design myself, the following passages were relatively meaningless to me. But later, having been enlightened to the validity of paradoxes, I was astonished when I read them as if for the first time. These remarkable scriptures display the paradoxical continuum very explicably:

- For in much wisdom is much grief: and he that increases knowledge increases sorrow (opposites are equally intense and equally learned: the more you learn of one, the more you learn of the other proportionately). (Ecclesiastes 1:18)

- That which has been (the past) is now (the present); and that which is to be (the future) has already been (opposites: the same thing except in reverse, with the peak and division in the midst). (Ecclesiastes 3:15)

- The thing that has been, it is that which shall be; and that which is done is that which shall be done: and there is no new thing under the sun (the continuum's unique cross-referencing learning curve: to know one is to know the other). (Ecclesiastes 1:9)

- To every thing (all) there is a season (one), and a time (one/season) to every purpose (all/everything) under the heaven (the perfect all in one and one in all organization). (Ecclesiastes 3:1)

The world still may not be ready for an advanced dual viewpoint. If not, this does not inhibit individuals from successfully applying it to their lives, and it will not be the first time or the last that some really good information was not in sync with the current, popular status quo. Regardless of an endorsement from the world of education, the benefits of this applicable, unique, and methodical study can not only assist us in independently conducting true scientific investigations, in spite of some of the bad information we are getting, but help us understand, guide, control, and confidently monitor the successful operating system for all the various departments in our personal lives with precision.

Because of these same canvassing tools that provide me with insight, I doubt very much that this independent research is going to set the world on fire, knowing that the greater the expert, the greater is the closure. However, I take great satisfaction in having my own confidence in these demonstrated laws and will continue their exploration, expecting more exciting discoveries in the future along with any good stuff the highly technical community has to offer. But I sincerely believe there are probably many individuals out there who are not bogged down with the personal and professional trappings that come with positions, having the ability and freedom to think on their own, that will benefit from understanding these astounding empirical truths. I also think there are some truth seekers that may have found the same or similar findings but got stumped in a few areas and couldn't make any breakthroughs. Perhaps many have kept silent because of the current political correctness madness or the strong opposition to the duality concept.

Therefore, it is my desire and aim to help make a logical connection for many thwarted independent thinkers out there or wannabes, assisting them in clearing the obscure scientific and political air by rolling back the mystic clouds, thus helping them to see the real world as it really is in real, mathematical, scientific terms, thereby fulfilling my dream and imaginative glory of one day hearing someone besides myself finally say, "It is possible to know what is truth and to obtain real answers."

Last, but far from least, I truly hope to exonerate this wonderful paradigm that is universally and innately embedded in all that we are and all that we see, the skeleton key (one) that unlocks any and every (all) door to volumes of connecting information, lifting it from its currently sad status of a blurry contradiction, educational stumbling block, and rejected cornerstone to its rightful place as the one and only quintessential summit of enlightenment and success. Optimistically and, I guess, a little dramatically, I wish to be able to shout from the rooftop one day that this well-kept secret is out, and what was considered to be the bad boy in town, ironically, turns out to be our shining hero and hidden treasure of true knowledge instead. And, finally, to experience the utmost pleasure in being able to share with others life's little magnificently complex, creative, DNA blueprint design that continually makes the world go around: the incredible, magical paradox.

Bibliography

ABC's of the Human Body, by *Reader's Digest*. The Reader's Digest Association, 1987.

Astronomy the Cosmic Journey, William K. Hartmann. Wadsworth Publishing Company, 1985.

Chaos, James Gleick. Viking, 1987.

"Classroom Assignment" Program on *Discovery Channel*. "Magnetism."

Discover, (October 2002). Excerpts from *The Blank Slate*, Steven Pinker.

Invitation to Philosophy, Issues and Options, by Stanley M. Honer and Thomas C. Hunt. Wadsworth Publishing Company, Fourth Edition, 1982.

The New Book Of Knowledge, The Children's Encyclopedia. "Atoms." Grolier, 1967.

The New Book Of Knowledge, The Children's Encyclopedia. "Philosophy." Grolier, 1967.

The New Book Of Knowledge, The Children's Encyclopedia. "Winds and Weather." Grolier, 1967.

Science and Invention Encyclopedia. Vol. 1, p. 12. H. S. Stuttman, 1987.

Science and Invention Encyclopedia. Vol. 1, p. 26. H. S. Stuttman, 1987.

Science and Invention Encyclopedia. Vol.1, p. 190. H. S. Stuttman, 1987.

Science and Invention Encyclopedia. Vol. 7, pp. 836–8. H. S. Stuttman, 1987.

Science and Invention Encyclopedia. Vol. 10, p.1373. H. S. Stuttman, 1987.

Science and Invention Encyclopedia. Vol. 18, pp. 2465–6. H. S. Stuttman, 1987.

Science and Invention Encyclopedia. Vol. 18, pp. 2467–8. H. S. Stuttman, 1987.

Smithsonian (October 2002). Book review by former editor, Paul Trachtman; *The Genius Within,* Frank T. Vertosick Jr.

The Tao of Physics, Fritjof Capra. Bantam Books, 1984.

Time (October 14, 2002). "Lacking in Self-Esteem? Good for You!" Andrew Sullivan.

World Book Encyclopedia. "Heart." CD-ROM.

World Book Encyclopedia. "Lungs." CD-ROM.

World Book Encyclopedia. "Magnetic Field." CD-ROM.

978-0-595-34480-2
0-595-34480-1

www.ingramcontent.com/pod-product-compliance
Lightning Source LLC
Chambersburg PA
CBHW030753180526
45163CB00003B/1002